Henry Albert Golding

The Theta-Phi Diagram

Practically Applied to Steam, Gas, Oil, and Air Engines

Henry Albert Golding

The Theta-Phi Diagram
Practically Applied to Steam, Gas, Oil, and Air Engines

ISBN/EAN: 9783743465145

Manufactured in Europe, USA, Canada, Australia, Japa

Cover: Foto ©berggeist007 / pixelio.de

Manufactured and distributed by brebook publishing software (www.brebook.com)

Henry Albert Golding

The Theta-Phi Diagram

THE
THETA-PHI DIAGRAM

PRACTICALLY APPLIED TO

STEAM, GAS, OIL, AND AIR ENGINES.

BY

HENRY A. GOLDING, A.M.I.M.E.,

Chief Draughtsman, Messrs. B. Donkin and Co., and Assistant Lecturer, South-Western Polytechnic, London.

PRICE THREE SHILLINGS NET.

1898.
THE TECHNICAL PUBLISHING CO. LIMITED,
31, WHITWORTH STREET, MANCHESTER.
JOHN HEYWOOD,
29 AND 30, SHOE LANE, LONDON; AND RIDGEFIELD, MANCHESTER.
And all Booksellers.

PREFACE.

In the following pages an attempt has been made to present in as simple and practical a manner as possible, the use of the temperature-entropy diagram and the various methods of drawing it for different heat motors. That the subject presented peculiar difficulties, because of its unfitness for presentation in a popular manner, will readily be granted; but I venture to think that one of the principal reasons for the lack of knowledge upon the subject by draughtsmen, steam students, and others has been the want of an elementary work, not overcrowded with mathematics. Most of the literature upon the subject has presented the mathematical rather than the graphical side of the question, with the result that students have become afraid of tackling what they believe to be an intricate mathematical investigation.

Of the utility of the temperature-entropy diagram in representing the various thermal changes which take place in all heat motors there cannot be any doubt. To quote only one authority, Mr. Mark H. Robinson, in the discussion on Mr. Willans' last paper, said: " Up to a certain point the practical man might ignore the present paper, and others like it; but if he aspired to design economical steam engines, he might derive more good from the study of, say, Mr. Macfarlane Gray's $\theta \phi$ diagram than from many portfolios of working drawings."

Where authorities have been quoted or made use of, the particulars are given in the text, but I will take this opportunity of expressing my indebtedness to Professor Ewing for his work on "The Steam Engine and other Heat Engines," and his Cantor Lectures on the "Mechanical Production of Cold"; to Professor Boulvin, for his articles in *La Revue de Mecanique;* and to various papers, principally those by the late Mr. P. W. Willans and Mr. Macfarlane Gray, published in the Proceedings of the Institutions of Civil and Mechanical Engineers. I also wish to thank the Council of the latter Institution for permission to reproduce some of the indicator diagrams and figures given in the reports of the Steam Jacket Research Committee.

PREFACE.

I regret that in Chapter IV., page 70, a slight inaccuracy should have occurred. Referring to the late Mr. Willans' method of calculating the thermal efficiency in his central valve engine tests, I find that the lower temperature of 110 deg. Fah. was only assumed for the preliminary calculations of the comparative losses due to incomplete expansion in condensing and non-condensing engines, but for the actual trials the temperature in the exhaust chamber was taken in each trial separately, and used for calculating the thermal efficiency. (See line 10 in Appendix, Table I., Willans on "Condensing Steam Engine Trials.") I am also indebted to Captain Sankey for pointing out that the specific heat of gases is not constant at the high temperatures occurring in gas and oil engines, and therefore the calculations which involve the use of Cp and Cv for a gaseous mixture at high temperatures must necessarily be looked upon as approximate only.

HENRY A. GOLDING.

London,
September, 1898.

CONTENTS.

	PAGE
INTRODUCTION	vii.

CHAPTER I.—ENTROPY.

ARTICLE
1. Introduction 1
2. Entropy Diagrams 2
3. Entropy 4
4. Theoretical Entropy Diagram for Steam 5

CHAPTER II.—ENTROPY OF WATER AND STEAM.

5. Method of Constructing the Curves 7
6. Entropy of Water 10
7. Entropy of Steam 11
8. Entropy Diagram for Ice, Water, and Steam 11
9. Constant Volume Curves 15

CHAPTER III.—CONVERSION OF INDICATOR DIAGRAM TO ENTROPY DIAGRAM.

10. Calculation of Dryness Fraction 1
11. Actual Example, Compound Engine 23
12. Complete Entropy Diagram (Professor Boulvin's Chart) ... 28
13. Application of Complete Entropy Diagram 33
14. $\theta \phi$ Diagram for Carnot Cycle 34
15. Condensation during Adiabatic Expansion 35
16. Adiabatic Expansion of Wet Steam 37

CHAPTER IV.—HEAT LOSSES.

17. Effect of Steam Jacketing 39
18. Theoretical Entropy Diagram for Superheated Steam ... 46
19. Effect of Superheating 49
20. Effect of Speed 56
21. Compounding 59
22. Initial Condensation 63
23. Measurement of Heat Losses 68

CONTENTS.

CHAPTER V.—Application to the Gas Engine.

	PAGE
24. General Considerations	73
25. Diagram for Theoretical Gas Engine	76
26. Diagram for Actual Gas-engine Trial	81
27. Corrected Diagram for Gas-engine Trial	89
28. Constant Volume Curves	92
29. Heat Losses in 7 Horse Power Gas Engine	94

CHAPTER VI.—Application to Oil and Air Engines.

30. Diagram for 20 Horse Power Diesel Motor	97
31. Stirling's Hot-air Engine	101
32. Ericsson's Hot-air Engine	104
33. Entropy Diagram for Refrigerators	105

APPENDIX.

Weight of Dry Saturated Steam	109

INTRODUCTION.

THE following contribution to the temperature-entropy method of graphically solving thermo-dynamic problems marks a further step in advance in the practical application of the system to the every-day questions that arise in the study of the steam engine and other heat motors. Although the method was foreshadowed by William Gibbs in 1873, it is only within the last few years that it has been applied in practice. Unfortunately, information respecting it is scattered about in the Proceedings of the Institution of Civil Engineers, in those of the Institution of Mechanical Engineers, and in various technical journals, both British and foreign. The Author has collected this information together, and has produced a work which treats the matter in a comprehensive manner, bringing it up to date so far as published materials allow.

Though not prepared to endorse every view expressed, I can fully recommend the book as likely to be eminently helpful to those studying the subject for the first time.

H. R. SANKEY, R.E.

THE ENTROPY DIAGRAM AND ITS APPLICATIONS.

CHAPTER I.—Entropy.

1. Introduction.

The representation of various forms of energy by means of a diagram has long been known and used with advantage by engineers and others. It is usually shown as a closed figure, the area of which represents energy (in either work, heat, or other units), and the ordinates, pressure or resistance overcome, and space passed through. The ordinary indicator diagram is perhaps the most common example of such figures, but it does not show the reception and distribution of heat which takes place in all steam and other heat engines. Considering the great advance made by the science of thermo-dynamics in recent years, it is somewhat surprising to find that the representation of the thermal changes which take place in all heat engines, in the form of a "heat diagram," has been so little applied for practical use. The relative advantages of the mathematical and graphical methods of representing the result of any process are so well known, that it will be unnecessary to refer to them here; beyond stating that where both methods can be employed the graphical very often becomes a useful adjunct to the mathematical, and is usually more easily grasped and understood by draughtsmen, students, and others.

The introduction of entropy diagrams is mainly due to Mr. J. Macfarlane Gray, who, in a paper read at the Paris meeting of the Institution of Mechanical Engineers in 1889, first showed the method of representing the heat contained by water, steam, and various ideal substances, on what he termed a ".theta-phi chart"; the vertical ordinates of which represented temperature, and the horizontal entropy. By denoting all absolute temperatures by the Greek letter θ (theta), and all quantities of entropy by the letter ϕ (phi), the diagram has come to be known in England as the "theta-phi diagram." In this book it will be preferable to adopt the Fahrenheit scale of temperature as that most used by practical men, and therefore absolute temperatures (Fah.) will usually be denoted by the Greek letter τ (tau), in accordance with most of the recent works on thermodynamics.

The practical application of the entropy diagram is perhaps more due to the late Mr. P. W. Willans, who, in a paper on "Non-condensing Steam-engine Trials," read before the Institution of Civil Engineers in 1888,[*] first used the diagram for the representation of steam-engine performance. Since then Capt. H. R. Sankey, R.E., has extended the subject by applying the diagram to the marine engines tested by the Research Committee of the Institution of Mechanical Engineers;[†] but there still seems to be a dearth of information on how to draw the diagrams that, it is hoped, the present book will, in a measure, supply.

2. Entropy Diagrams.

The theta-phi or temperature-entropy diagram is a graphic representation of the thermal changes which take place in a steam-engine cylinder during one cycle. It is plotted on a theta-phi chart, by calculations made from the mean indicator diagram of an experiment. As an aid to the thermo-dynamic study of the steam engine, it is much more useful than the better known indicator diagram, as it shows at a glance the thermal efficiency of the engine. The ordinary indicator

---

[*] See Proc. Inst. of Civil Engineers, vol. xciii., Part III.
[†] See Proc. Inst. of Mechanical Engineers, February, 1894.

diagram only shows the amount of work done, independent of the amount of steam used; but the theta-phi (or $\theta\phi$ as it is better written) diagram shows, by its area, the proportion of heat utilised to the heat received.

COMPARISON OF $\theta\phi$ DIAGRAM WITH INDICATOR DIAGRAM.

In the ordinary indicator diagram the area represents the work done on the piston in one stroke, the vertical ordinates being "pressure," and the horizontal abscissæ "distance moved through." Similarly, in the $\theta\phi$ diagram, the area

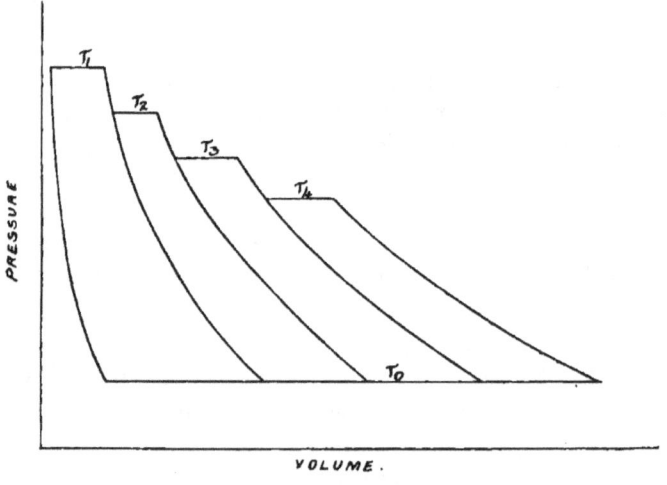

FIG. 1.

also represents work done, but in *heat units;* the vertical ordinates being absolute temperature (denoted by the Greek letter θ), and the horizontal dimension is what Clausius termed "entropy," named by Zeuner "heat weight," and represented by the Greek letter ϕ. In the indicator diagram the area represents the product of pounds (pressure) × feet (distance), or foot-pounds (work); so in the $\theta\phi$ diagram the area represents the product of θ (absolute temperature) × ϕ (entropy) in thermal units (work).

3. Entropy.

Entropy is the co-ordinate with temperature of energy; *i.e.*, it is the length upon a diagram whose height is absolute temperature, and whose area is heat units.

The meaning of the term will be better understood in the light of Carnot's principle of efficiency applied to a reversible cycle in which the heat is received at various temperatures. Fig. 1 represents such a cycle, where, in the first stage, heat is received at τ_1 and discharged at τ_0; in the second stage, a further quantity of heat is received at τ_2 and discharged at τ_0, and so on. Let Q_1, Q_2, Q_3, &c., represent the various quantities of heat received at each stage. The area of the whole figure, which represents in heat units the work done W, will equal

$$W = \frac{Q_1}{\tau_1}(\tau_1 - \tau_0) + \frac{Q_2}{\tau_2}(\tau_2 - \tau_0) + \frac{Q_3}{\tau_3}(\tau_3 - \tau_0) + \ldots$$

or, for a general formula,

$$W = \frac{Q}{\tau}(\Delta \tau), \text{ in heat units;}$$

where Q = heat received,
  τ = temperature of reception (absolute),
  $\Delta \tau$ = difference of temperature between reception and rejection of heat.

Let R = the heat rejected to the cold body,
  then $Q = W + R$;
but $Q = Q_1 + Q_2 + Q_3 + \ldots$
and $W = \frac{Q_1}{\tau_1}(\tau_1 - \tau_0) + \frac{Q_2}{\tau_2}(\tau_2 - \tau_0) + \frac{Q_3}{\tau_3}(\tau_3 - \tau_0) + \ldots$

or, by difference, $R = \frac{Q_1 \tau_0}{\tau_1} + \frac{Q_2 \tau_0}{\tau_2} + \frac{Q_3 \tau_0}{\tau_3} + \ldots$

or, $\frac{R}{\tau_0} = \frac{Q_1}{\tau_1} + \frac{Q_2}{\tau_2} + \frac{Q_3}{\tau_3} + \ldots$

or, $\frac{Q_1}{\tau_1} + \frac{Q_2}{\tau_2} + \frac{Q_3}{\tau_3} + \ldots - \frac{R}{\tau_0} = 0.$

or (the sum) $\Sigma \frac{Q}{\tau} = 0.$

ENTROPY. 5

That is to say, the algebraical sum of the changes of entropy in any complete reversible cycle is *nil*; therefore the entropy diagram for any reversible cycle must be a closed figure; and the final quantity of entropy will be equal to the initial, no matter where the process be started. Entropy, therefore, is the fraction or ratio $\frac{Q}{\tau}$, representing the amount of heat taken up or rejected by a body, divided by its absolute temperature at that time. It can be calculated from any arbitrary zero of temperature; for water and steam, it will be found most convenient to calculate the entropy above that already possessed by 1 lb. of water at 32 deg. Fah., so as to avoid including the latent heat of water. It is usually plotted with temperature as ordinates to an abscissa of entropy.

4. THEORETICAL ENTROPY DIAGRAM FOR STEAM.

The thermal changes which take place when water is evaporated, expanded (as steam), and condensed, are represented on the $\theta \phi$ diagram in the following way: Take 1 lb. of water at the condenser temperature, say τ_3 deg. Fah. absolute, and let A represent its position on the diagram, fig. 2, its temperature τ_3 and entropy O a being known. Now pump it into the boiler, where it is heated from τ_3 to τ_1 deg. Fah., and, as its temperature is increased, the process will be shown by an upward curve A F B, to correspond with the vertical temperature scale; but, as it also receives heat, the curve must progress to the right to indicate its increased entropy. The change in its state by heating it from τ_3 to τ_1 deg. is therefore shown by the curve A F B, every increment of temperature being accompanied by a corresponding increase in entropy, denoted by $\frac{h}{\tau}$, where h is the heat contained in 1 lb. of water at τ deg. absolute temperature. Having reached the temperature τ_1 of the water in the boiler, it begins to evaporate, its temperature remains constant, but it receives an amount of heat (L_1) known as the latent heat of 1 lb. of steam at τ_1 deg. temperature. This is represented in fig. 2 by the horizontal line B C—horizontal, because its temperature does not change during the operation; the amount bc of its increase in entropy

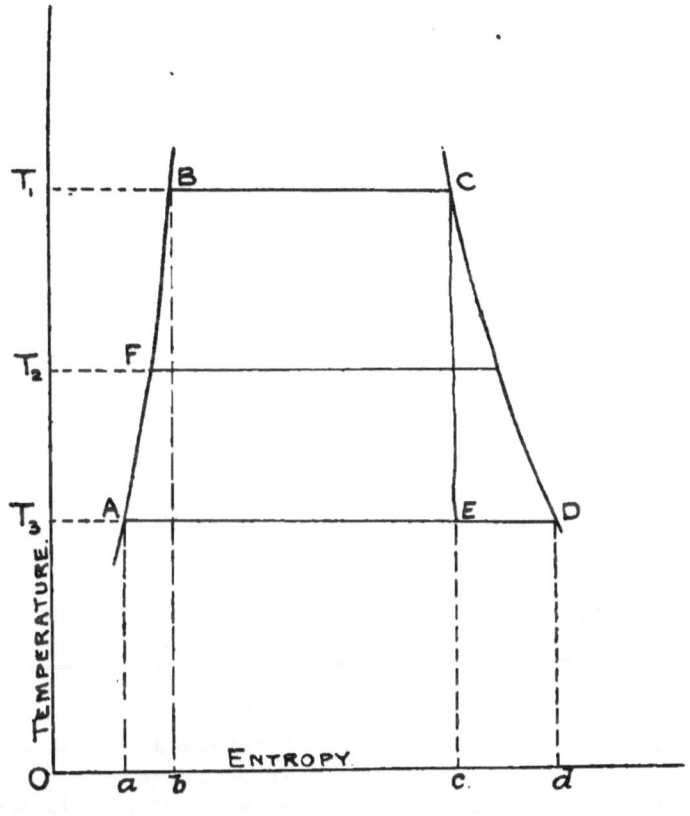

Fig. 2.

being equal to its latent heat divided by its absolute temperature, or $\frac{L_1}{\tau_1}$. Its total entropy Oc is made up of the two quantities Ob and bc, representing the heat of formation as water (usually denoted by h), and the latent heat L, respectively; its total heat H being the sum of the

ENTROPY OF WATER AND STEAM. 7

two, or $h + L$. At C the steam is admitted into the cylinder, and allowed to expand. If external heat be added to it during expansion, so as to keep it up to saturation point, it will follow the law of the saturation curve $pv^{1.0645} = $ a constant, and its entropy will be denoted by the curve C D, such that all horizontal dimensions from A B to C D are *equal* to the latent heat of 1 lb. of dry saturated steam divided by its absolute temperature. If the steam expands adiabatically, the entropy curve will be a straight vertical line C E ; because, as the steam neither receives nor loses heat, its *entropy* will be unchanged. This also shows clearly the amount of *wetness* which always accompanies adiabatic expansion, and the ratio of A E to A D represents the dryness fraction of the steam at the end of expansion. To indicate the condensing operation, the curve returns along the horizontal line from D to A, the temperature of the mixture remaining constant at τ_3 deg., and its entropy being reduced from O d to O a.

CHAPTER II.—ENTROPY OF WATER AND STEAM.

5. METHOD OF CONSTRUCTING THE CURVES.

HAVING explained the operations to which the theoretical entropy diagram for steam refers, it is necessary to find a means of drawing the two boundary curves A B and C D of the $\theta \phi$ chart shown in fig. 2. For ordinary steam engines it is not necessary to refer to temperatures below 100 deg. Fah. (corresponding to a pressure of about 1 lb. absolute), nor higher than 400 deg. Fah., equal to about 260 lb. absolute pressure. We have therefore 300 deg. Fah. range of temperature to provide for, and a scale of 20 deg. Fah. to 1 in. will be found convenient if the chart be drawn on an ordinary sheet of sectional paper. For the base line, or entropy, starting from water at 100 deg. Fah., the maximum required will be 1·87 ; so that an entropy scale of 0·1 ϕ = 1 in. will be ample. These scales will give for area, 1 square inch =

8   THE ENTROPY DIAGRAM AND ITS APPLICATIONS.

20 deg. $\theta \times 0{\cdot}1\,\phi = 2{\cdot}0$ British thermal units. It must be distinctly understood that the $\theta\phi$ diagram is always drawn for 1 lb. of H_2O, whether it be steam, water, or a mixture of both steam and water. Having plotted the scales, we start at the bottom left-hand corner of the chart with 1 lb. of

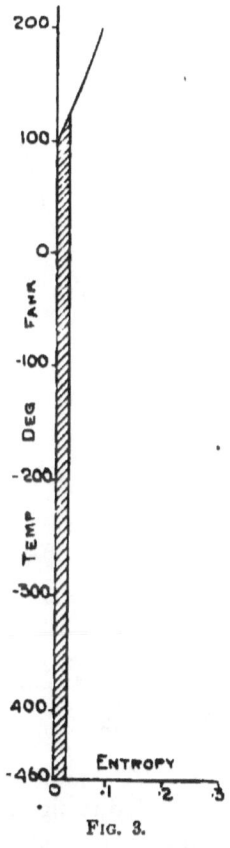

Fig. 3.

water at 100 deg. Fah., or 560 deg. Fah. absolute, and calculate the quantities of heat as above. First construct the curve of entropy of water, or *aqueus curve* as it is sometimes called, shown by A B, in fig. 2. As this curve is almost a straight line it is only necessary to calculate the value of entropy for some 10 or 12 points, so commencing with water heated from 100 deg. Fah. to 125 deg. Fah. (see fig. 3), the heat given to 1 lb. of water to raise its temperature from 560 deg. Fah. absolute to 585 deg. will be represented on the

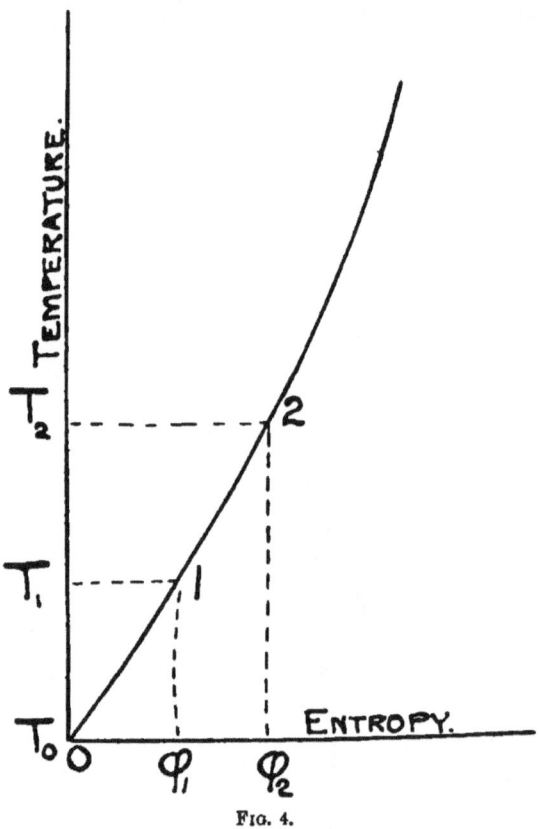

Fig. 4.

$\theta \phi$ diagram by the area shaded in fig. 3. The area down to the absolute zero of temperature must be added, so as to include all the heat contained in the water. In this case, the heat received is 25·08 B.T.U. (see tables of Properties of Saturated Steam), and, therefore, the increase of entropy will be $\frac{25\cdot08}{572\cdot5}$, or 0·0438. Similarly, the value of ϕ can be calculated for any temperature, and should be tabulated as in Table I., both for the purpose of plotting the aquene curve, and for future reference. Table I. gives the values of the entropy of 1 lb. of water for every 10 deg. Fah. from 32 deg. Fah. to 400 deg. Fah.

6. Entropy of Water.

The aquene curve can also be plotted from calculations made by the aid of the calculus in the following manner: For any increase of temperature from τ_0 to τ_1 (see fig. 4),

$$\phi_1 = \frac{\Delta h}{\tau},$$

where Δh represents the heat necessary to raise 1 lb. of water from τ_0 to τ_1; and τ is the mean absolute temperature during the operation. Integrating this, we get—

$$\phi_1 = \int_{\tau_0}^{\tau_1} \frac{dh}{\tau};$$

and, assuming the specific heat of water as unity,

$$dh = d\tau;$$

or

$$\phi_1 = \int_{\tau_0}^{\tau_1} \frac{dt}{\tau};$$

and, solving this, we get

$$\phi_1 = \log_e \tau_1 - \log_e \tau_0;$$

or

$$\phi_1 = \log_e \frac{\tau_1}{\tau_0};$$

that is, $d\phi$, or any small difference in the entropy of 1 lb. of water at any two temperatures is equal to the difference of the hyperbolic logarithms of the absolute temperatures. If the result be multiplied by the mean specific heat of water between the temperatures τ_1 and τ_0, the formula becomes

$$\phi_{\tau_1} - \phi_{\tau_0} = s (\log_e \tau_1 - \log_e \tau_0),$$

where s represents the heat necessary to raise 1 lb. of water 1 deg. Fah., between τ_0 and τ_1, as compared with the heat required to raise 1 lb. of water from 39 deg. Fah. to 40 deg. Fah. The values of s are given in Table I., page 12, together with the increase of entropy for every 10 deg. Fah., calculated by the above formulæ, and the total entropy above water at 32 deg. Fah. The last column in Table I. gives the difference of entropy per 1 deg. Fah., for the purpose of interpolation.

7. Entropy of Steam.

To draw the entropy curve for steam (C D, in fig. 2), an amount of entropy equal to $\frac{L}{\tau}$ must be added to the aqueus curve, where L is the latent heat of 1 lb. of steam at τ deg. absolute temperature. The values of L and $\frac{L}{\tau}$ are given in Table II., page 13, for every 10 deg. Fah. from 32 deg. to 400 deg., together with the difference of entropy of steam per 1 deg. Fah. for interpolating, and the total entropy of steam and water (above 32 deg. Fah.) with its difference per 1 deg. It should be noted that ϕ_{w+s}, the entropy of water and steam at any temperature, is not equal to $\frac{H}{\tau}$, where H = total heat of evaporation from 32 deg. Fah. at absolute temperature τ; because h, the sensible heat of the water, is not all received at temperature τ, but at a gradually increasing temperature.

8. Entropy Diagram for Ice, Water, and Steam.

The entropy diagram for ice, water, and steam is shown in fig. 5, page 14. The curve A B, for ice, is drawn on the assumption that its specific heat is 0·504 at all temperatures. The heat given to 1 lb. of ice per 1 deg. Fah. rise of temperature is, therefore, equal to 0·504 B.T.U.,

or
$$h = 0·504 \, dt;$$

$$\therefore \frac{h}{\tau} = \phi = 0·504 \frac{dt}{\tau};$$

or
$$\phi_2 - \phi_1 = 0·504 \times \log_e \frac{\tau_2}{\tau_1}.$$

TABLE I.

Entropy of water, from 32 deg. Fah. to 400 deg. Fah., calculated from the formulæ, $\phi_{\tau_1} - \phi_{\tau_0} = s\,(\log_e \tau_1 - \log_e \tau_0)$.

Temperature. Deg. Fah. t.	Absolute temperature. τ.	Specific heat. s.	Increase of entropy per 10 deg. Fah. $\phi_{\tau_1} - \phi_{\tau_0}$.	Total entropy above water at 32 deg. Fah. ϕ_w.	Difference of entropy per 1 deg. Fah. $\Delta \phi_w$.
32	492	..	....	....	0·00202
40	500	1·0	0·01613	0·01613	0·00198
50	510	1·0	0·01980	0·03593	0·00194
60	520	1·0	0·01942	0·05535	0·00191
70	530	1·001	0·01906	0·07441	0·00187
80	540	1·001	0·01871	0·09312	0·00184
90	550	1·002	0.01838	0·11150	0·00181
100	560	1·002	0·01806	0·12956	0·00177
110	570	1·003	0·01775	0·14731	0·00174
120	580	1·004	0·01745	0·16476	0·00172
130	590	1·004	0·01716	0·18192	0·00169
140	600	1·005	0·01689	0·19881	0·00166
150	610	1·006	0·01662	0·21543	0·00164
160	620	1·007	0·01637	0·23180	0·00161
170	630	1·008	0·01613	0·24793	0·00159
180	640	1·009	0·01589	0·26382	0·00157
190	650	1·010	0·01566	0·27948	0·00154
200	660	1·011	0·01544	0·29492	0·00152
210	670	1·012	0·01522	0·31014	0·00150
220	680	1·013	0·01501	0·32515	0·00148
230	690	1·014	0·01480	0·33995	0·00146
240	700	1·015	0·01461	0·35456	0·00144
250	710	1·017	0·01443	0·36899	0·00142
260	720	1·019	0·01425	0·38324	0·00141
270	730	1·021	0·01408	0·39732	0·00139
280	740	1·022	0·01391	0·41123	0·00137
290	750	1·024	0·01374	0·42497	0·00136
300	760	1·026	0·01358	0·43855	0·00134
310	770	1·027	0·01343	0·45198	0·00133
320	780	1·029	0·01328	0·46526	0·00131
330	790	1·031	0·01313	0·47839	0·00130
340	800	1·032	0·01298	0·49137	0·00128
350	810	1·034	0·01284	0·50421	0·00127
360	820	1·036	0·01271	0·51692	0·00126
370	830	1·038	0·01258	0·52950	0·00124
380	840	1·040	0·01245	0·54195	0·00123
390	850	1·042	0·01233	0·55428	0·00122
400	860	1·044	0·01221	0·56649	

TABLE II.

Entropy of Dry Saturated Steam, **from 32 deg. Fah. to 400 deg. Fah.**

Temperature. Deg. Fah.	Absolute temperature.	Latent heat.	Entropy of 1 lb. steam.	Difference of entropy per 1 deg. Fah.	Total entropy of water + steam.	
					Per lb.	Difference per 1 deg. Fah.
t.	τ.	L.	$\phi_s = \dfrac{L}{\tau}$.	$\Delta\, \phi_s$.	$\phi_w + \phi_s$.	$\Delta\, \phi w + s$.
32	492	1091·7	2·2189	0·00581	2·2189	0·00380
40	500	1086·2	2·1724	0·00561	2·1885	0·00363
50	510	1079·3	2·1163	0·00542	2·1522	0·00348
60	520	1072·3	2·0621	0·00521	2·1174	0·00330
70	530	1065·3	2·0100	0·00502	2·0844	0·00315
80	540	1058·3	1·9598	0·00483	2·0529	0·00299
90	550	1051·3	1·9115	0·00465	2·0230	0·00285
100	560	1044·36	1·8649	0·00449	1·9945	0·00272
110	570	1037·39	1·8200	0·00434	1·9673	0·00259
120	580	1030·42	1·7766	0·00420	1·9414	0·00249
130	590	1023·40	1·7346	0·00406	1·9165	0·00237
140	600	1016·39	1·6940	0·00393	1·8928	0·00227
150	610	1009·38	1·6547	0·00379	1·8701	0·00216
160	620	1002·37	1·6167	0·00368	1·8485	0·00207
170	630	995·33	1·5799	0·00357	1·8278	0·00198
180	640	988·30	1·5442	0·00346	1·8080	0·00189
190	650	981·24	1·5096	0·00336	1·7891	0·00182
200	660	974·18	1·4760	0·00326	1·7709	0·00174
210	670	967·10	1·4434	0·00316	1·7535	0·00166
220	680	960·03	1·4118	0·00307	1·7369	0·00158
230	690	952·94	1·3811	0·00299	1·7211	0·00153
240	700	945·83	1·3512	0·00292	1·7058	0·00148
250	710	938·76	1·3220	0·00282	1·6910	0·00140
260	720	931·56	1·2938	0·00275	1·6770	0·00134
270	730	924·41	1·2663	0·00268	1·6636	0·00129
280	740	917·25	1·2395	0·00261	1·6507	0·00123
290	750	910·06	1·2134	0·00254	1·6384	0·00119
300	760	902·86	1·1880	0·00248	1·6265	0·00113
310	770	895·64	1·1632	0·00242	1·6152	0·00109
320	780	888·41	1·1390	0·00236	1·6043	0·00105
330	790	881·15	1·1154	0·00231	1·5938	0·00101
340	800	873·88	1·0923	0·00225	1·5837	0·00097
350	810	866·58	1·0698	0·00219	1·5740	0·00092
360	820	859·27	1·0479	0·00215	1·5648	0·00089
370	830	851·95	1·0264	0·00210	1·5559	0·00085
380	840	844·58	1·0054	0·00205	1·5474	0·00082
390	850	837·20	0·9849	0·00200	1·5392	0·00078
400	860	829·84	0·9649		1·5314	

Thus ϕ becomes infinity when $\tau = 0$; and the real origin of the entropy scale should be at infinity on the left-hand side of the figure. For convenience of plotting the origin

14   THE ENTROPY DIAGRAM AND ITS APPLICATIONS.

has been taken as for water at 492 deg. Fah. (absolute), the amount of entropy on either side of this point being considered as positive and negative quantities. The break in the curve between B and C is due to the latent heat of water, or the heat required by 1 lb. of ice at 492 deg. for

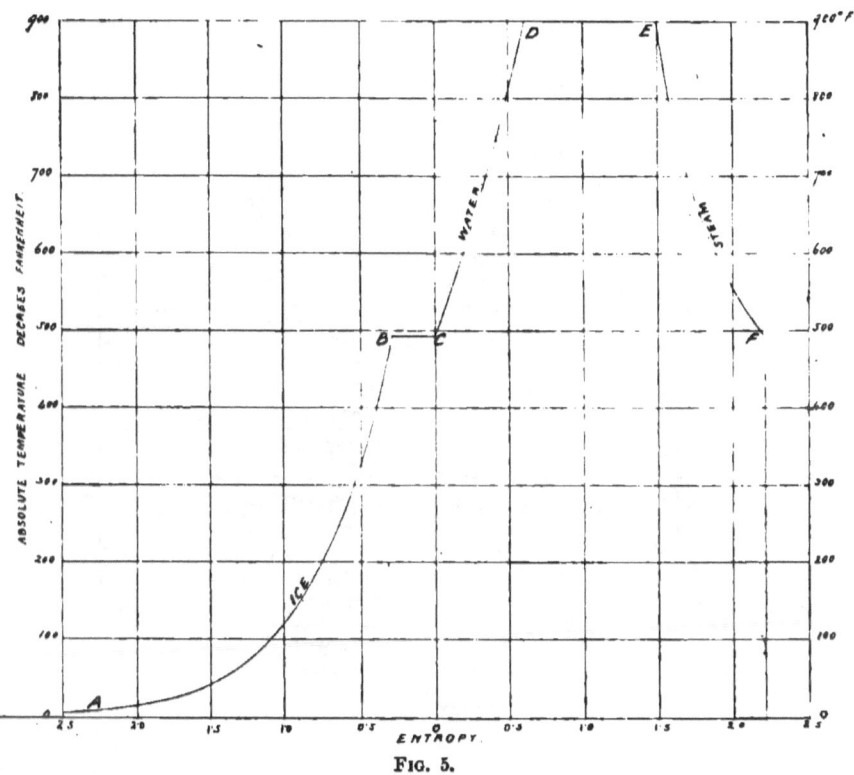

Fig. 5.

conversion into water at 492 deg. Its amount is equal to 143 B.T.U., and its entropy is therefore equal to $\frac{143}{492} = 0.29$.

The curves for water and steam are drawn as already explained. If the two curves C D and F E be produced, they will meet at a point known as the "critical temperature" for water and steam, above which there can be no liquid. Mr. Macfarlane Gray* finds this to be at about 750 deg. Cen., or about 1,350 deg. Fah. (absolute). The area of the diagram, fig. 5, is such that each square represents 100 × 0·5 = 50 B.T.U.

9. CONSTANT VOLUME CURVES.

Having constructed the two boundary curves of the $\theta\phi$ chart (the most tedious and difficult part of the process), it will be advisable to draw "constant pressure lines" and "con-

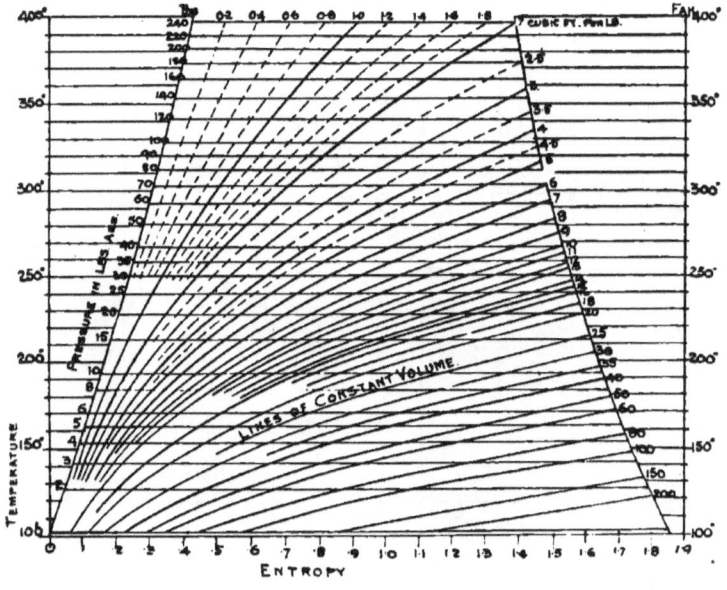

FIG. 6.

stant volume curves," as shown in fig. 6, in order to facilitate the transfer of indicator diagrams. The former are simply horizontal lines drawn across the chart between the two boundary curves, at a height corresponding to the *temperature* of dry saturated steam, as given in the steam tables. If a temperature scale of 20 deg. Fah. to 1 in. be adopted, the

---

* See Proc. Inst. of Mechanical Engineers, July, 1889, page 419.

pressure lines may be drawn for every 1 lb. pressure up to about 20 lb. absolute, but beyond this they should be reduced to every 2 lb., and afterwards to 5 lb. intervals, to prevent complicating the chart. These constant pressure lines are shown in fig. 6.

The constant volume curves present a little more difficulty. They represent the loss of entropy due to a drop in pressure at constant volume, or that which takes place in the ordinary engine cylinder at release, when it opens to the condenser.

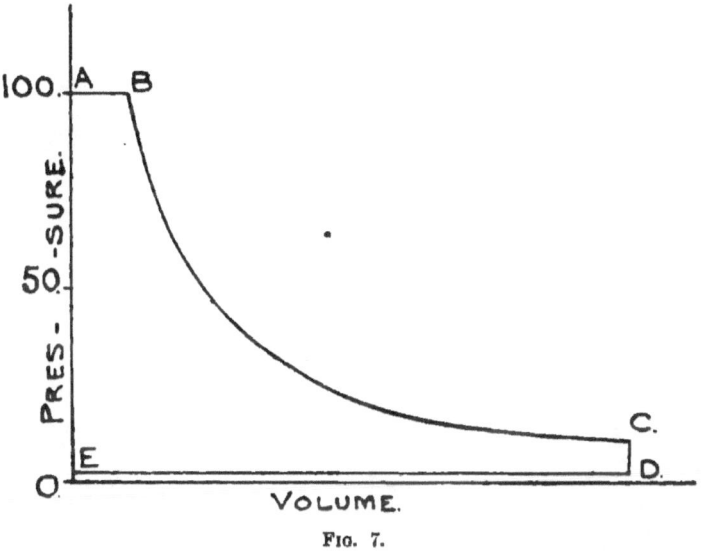

Fig. 7.

For example, suppose 1 lb. of steam at 100 lb. pressure to expand, keeping dry, to 10 lb. absolute, and then to open to the condenser so that the pressure falls to 2 lb. absolute at constant volume, as shown in fig. 7. The 1 lb. of dry steam at 10 lb. pressure occupies 37·8 cubic feet, and has an entropy of 1·64 above water at 100 deg. Fah.; its pressure falls to 2 lb. by condensing, without any increase in its volume, thus causing a reduction of entropy in the steam. At 2 lb. pressure it would occupy 173 cubic feet if it were

CONSTANT VOLUME CURVES. 17

all steam; but we know that it only occupies 37·8 cubic feet, and therefore there can only be $\frac{37\cdot8}{173}$, or 0·218 lb. of steam in the cylinder after opening to the condenser, which will possess an entropy represented by the distance E D, in fig. 8, where

$$\frac{ED}{EF} = 0\cdot218;$$

the curve C D forming a part of one of the constant volume curved lines which have to be drawn. To construct these

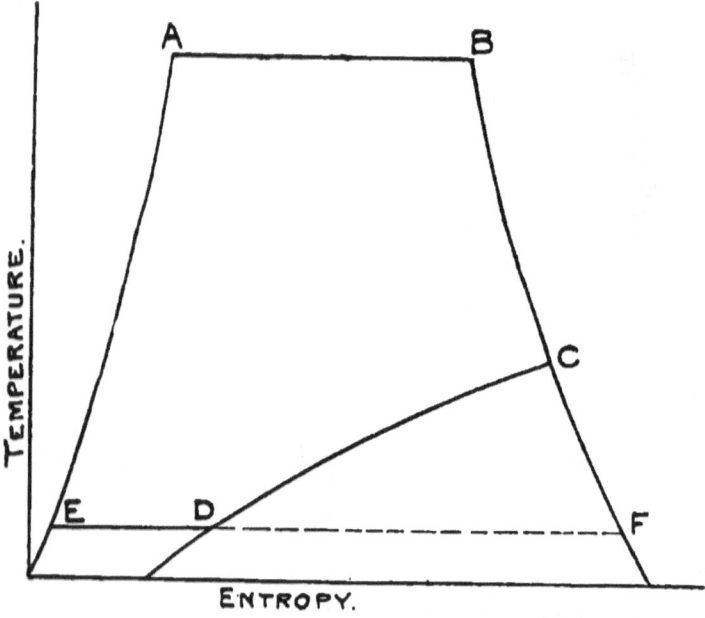

FIG. 8.

curves, divide the horizontal distance or entropy between the water and steam curves, at any particular pressure, into as many equal parts as there are cubic feet of volume to 1 lb. of dry saturated steam at that pressure. For example, at 60 lb. absolute pressure the specific volume of steam is practically 7 cubic feet; and therefore the curves for 1, 2, 3,

3T

18   THE ENTROPY DIAGRAM AND ITS APPLICATIONS.

4, 5, and 6 cubic feet constant volume lines (fig. 6) all cut the 60 lb. pressure line at equal horizontal distances from one another.

Another method of drawing the constant volume curves is based on the principle that

$$\frac{d\,P}{d\,\tau} \text{ varies as } \frac{L}{\tau}, \text{ or varies as } \phi.$$

This is found by equating the work done to the heat expended in a perfect steam engine working on the Carnot

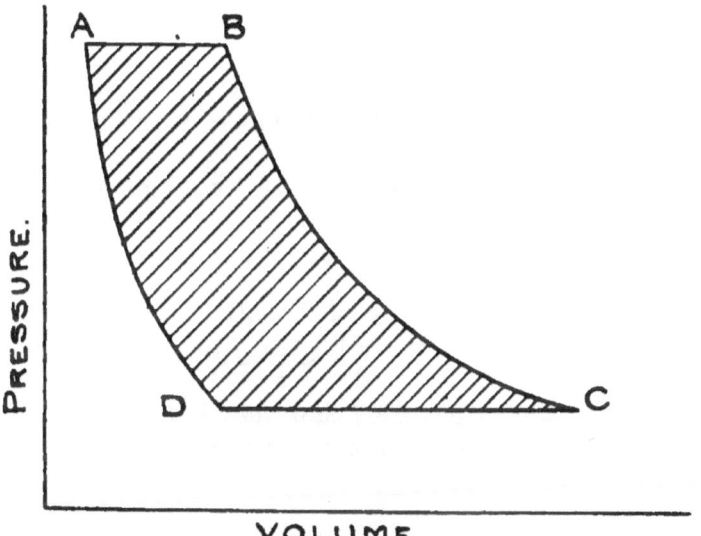

Fig. 9.

cycle. Take 1 lb. of water of volume O A, fig. 9, and pressure P_1; evaporate it to dry steam of volume O B and temperature τ_1; expand it adiabatically to C, fig. 9, where its pressure is P_2, and temperature τ_2; now condense it at this pressure and temperature until it occupies the volume O D; and, finally, complete the cycle by compressing it adiabatically until it occupies its initial volume of O A.

The efficiency of the cycle will be $\frac{\tau_1 - \tau_2}{\tau_1}$; and the work done will be equal to the area shaded in fig. 9,

$$= L\left(\frac{\tau_1 - \tau_2}{\tau_1}\right)$$

in heat units, or $J \cdot L\left(\frac{\tau_1 - \tau_2}{\tau_1}\right)$ in work units. Now, if we make $P_1 - P_2 = dP$ very small, and $\tau_1 - \tau_2 = d\tau$ also very small, the work done will be equal to the rectangle A B C D, which equals

$$(\text{vol. O B} - \text{vol. O A}) \cdot (P_1 - P_2)$$
$$= (V - v) \, dP,$$

where V and v are the volumes of 1 lb. of steam and 1 lb. of water respectively.

But the work done is equal to the heat expended; therefore

$$(V - v) \cdot dP = J \cdot L\left(\frac{d\tau}{\tau_1}\right)$$

Transposing J, and dividing all through by $d\tau$, we get—

$$\frac{V - v}{J} \times \frac{dP}{d\tau} = \frac{L}{\tau} = \phi.$$

But $v =$ approximately 0·017 cubic feet, and V is a constant for any pressure, and J a numerical constant; therefore, for any particular pressure,

$$\frac{dP}{d\tau} \text{ varies as } \frac{L}{\tau}, \text{ or varies as } \phi.$$

In other words, the horizontal ordinates of any constant volume curve at various pressures are proportional to the value of $\frac{dP}{d\tau}$ for those particular pressures. For example, take 1 lb. dry steam at 150 lb. absolute pressure, which has a specific volume of 2·978 cubic feet. The value of

$$\frac{dP}{d\tau} = 1\cdot 91$$

(see Table I$_A$., Appendix of Cotterill on the "Steam Engine"), and $\frac{L}{\tau}$ or ϕ for steam alone (see Table II., page 13)

$$= \frac{860\cdot 62}{818\cdot 16} = 1\cdot 052.$$

At 100 lb. pressure,

$$\frac{d\,P}{d\,\tau} = 1\cdot39,$$

and therefore the ϕ ordinate for the 2·978 constant volume curve will be

$$\frac{1\cdot39}{1\cdot91} \times 1\cdot052 = 0\cdot765.$$

The values of ϕ for 2·978 cubic feet of steam calculated in this way are given in Table III., and if they are plotted on the $\theta\,\phi$ chart at their proper pressures, they will be found to form a constant volume curve for 2·978 cubic feet, which will be just inside the 3 cubic feet curve already drawn by the geometrical method previously explained.

TABLE III.

Value of Entropy for 2·978 cubic feet of steam, for drawing constant volume line.

Absolute pressure. Pounds per square inch. p	Temperature Fahrenheit. Degrees. t	Ratio of $\dfrac{d\,p}{d\,t}$ from steam tables.	Entropy of 2·978 cubic feet of steam, by proportion.
150	358·16	1·91	1·05
100	327·57	1·39	0·765
50	280·85	0·79	0·435
20	227·92	0·376	0·207
10	193·24	0·213	0·117
5	162·33	0·119	0·065
2	126·27	0·055	0·030

CHAPTER III.—CONVERSION OF INDICATOR DIAGRAM TO
ENTROPY DIAGRAM.

10. CALCULATION OF DRYNESS FRACTION.

To convert the ordinary indicator diagram to the $\theta\phi$ diagram, it will be necessary to have the following data furnished by an experiment :—

(*a*) Exact size of cylinders, and clearance volumes of each end of each cylinder.

(*b*) Dryness fraction of the steam in each cylinder at any one period during expansion.

(*c*) A mean indicator diagram for each cylinder.

The diameters of the cylinders should be taken from gauges, and corrected to allow for the expansion of the cylinders when hot. The clearance volumes can be obtained either by calculation or direct measurement by filling the ports, &c., with water. When the latter method is employed, care must be taken to allow for the escape of the air ; and where possible, it is advisable to calculate the volume as well as measure it, the one method serving as a check upon the other.

The dryness fraction is calculated from the mean indicator diagram, when the weight of steam used by the engine per stroke is known, in the following manner : Take any point in the expansion period of the stroke, as A, fig. 10, and scale the pressure P_a shown by the indicator diagrams at that point. Calculate the volume (V_a) occupied by the steam in the cylinder at that portion of the stroke ; *i.e.*, if A be taken at half stroke,

$$V_a = \frac{A \times L}{2} + v,$$

where A = mean area of front and back of piston in square feet ;

L = length of stroke in feet ;

v = mean clearance volume of front and back in cubic feet.

22 CALCULATION OF DRYNESS FRACTION.

Multiply the volume V_a by the weight of 1 cubic foot of steam at P_a pounds absolute pressure, and obtain the weight (W_a) of steam per stroke shown to be present in the cylinder at point A. Subtract from this weight (W_a), the weight of steam (W_b) left in the cylinder from the previous stroke at the end of compression, as shown at B, fig. 10. This will be equal to V_b × density of steam at P_b pounds absolute pressure; and the *net* weight of steam per stroke accounted for by the indicator at point A will be $W_a - W_b$ pounds. Compare this with the net weight of steam used by the engine per stroke, which can be obtained by measuring

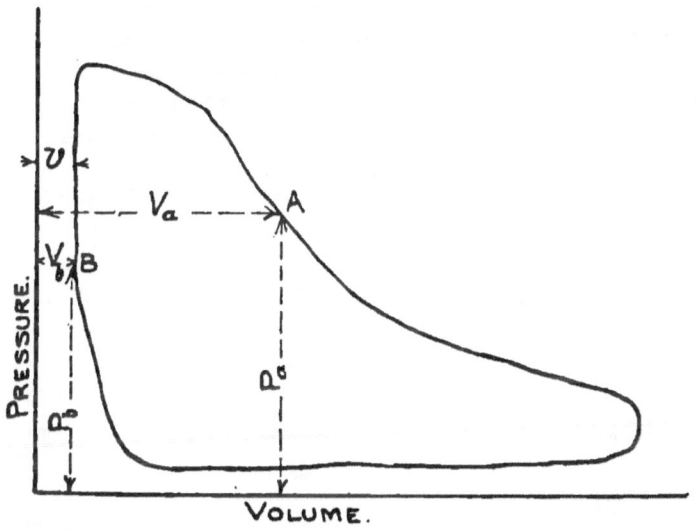

FIG. 10.

the feed water and allowing for steam pipe and jacket condensation, or by measuring the water discharged by the air pump if a surface condenser be used, and the proportion of $\dfrac{\text{indicated steam}}{\text{steam used}}$ is the required dryness fraction at point A. A table of the weight of dry saturated steam for various pressures will be found in the appendix.

COMPOUND ENGINE AT HAMPTON. 23

To obtain an average indicator diagram for each cylinder, all the diagrams taken during the trial should be divided into 20 parts, and the forward and backward pressures marked off on long strips of paper, the means so obtained being tabulated as shown in Table V., page 25, together with the volume of the cylinder plus clearance at that part of the stroke.

11. ACTUAL EXAMPLE: COMPOUND ENGINE AT HAMPTON.

As an example, take the trial of a compound pumping engine at Hampton, of the Southwark and Vauxhall Waterworks, and tested by Prof. Hudson Beare in January, 1894. Full particulars of the trial will be found in the third report of the Steam Jacket Research Committee of the Institution of Mechanical Engineers, published in October, 1894. The engine is of the vertical, inverted, surface-condensing type,

TABLE IV.—TRIAL OF COMPOUND ENGINE AT HAMPTON.
Data from Proc.Inst.Mech.E., October, 1994.

	H.P. cylinder.	L.P. cylinder.
Total volume of cylinder .... cubic feet	38·41	105·05
Volume of cylinder at release .... cubic feet	36·49	99·80
Volume of clearance .... cubic feet	0·96	1·94
Total volume of steam at release .... cubic feet	37·35	101·74
Pressure at release .... lbs. absolute p_1	26·12	8·02
Density of steam at above pressure .... w_1	0·06515	0·02145
Weight of steam present at release .... W lbs.	2·4332	2·1822
Pressure at end of compression .... lbs. absolute p_2	80·1	12·42
Density of steam at above pressure .... w_2	0·18645	0·03239
Weight of steam left in clearance .... w lbs.	0·1609	0·0628
Net steam accounted for by indicator .... (W – w) lbs.	2·2723	2·1194
Dryness fraction at release .... per cent	92·1	85·9

with cylinders 32 in. and 52⅝ in. diameter; stroke, 7 ft.; piston rods, 6 in. diameter. Considering only the trial with all the steam jackets on, made on January 4th, 1894, Table LX., page 578 of the report, gives the particulars from which the figures given in Table IV. are taken. From the mean indicator diagrams published in plate 143 of the committee's report, the pressures at every one-twentieth part of the stroke in both cylinders have been scaled, and tabulated in

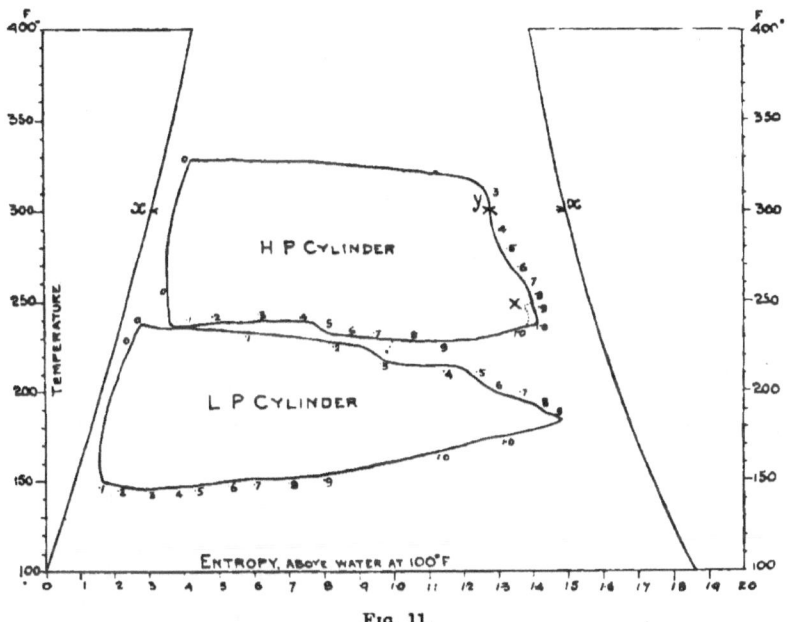

Fig. 11.

Tables V. and VI. To calculate the $\theta\phi$ volume—that is, the volume which 1 lb. of steam of the required dryness would occupy—any initial point may be taken where the dryness fraction of the steam is known; as at release, or 95 per cent of the forward stroke. In this particular trial, the weight of steam used was 6,284 lb. per hour, or 2·4665 lb. per stroke of steam and water passing through the cylinders, as found by measuring the air-pump discharge; and at release, in the high-pressure cylinder, this weight of steam occupies 37·35

COMPOUND ENGINE AT HAMPTON. 25

TABLE V.—$\theta\phi$ DIAGRAM FOR COMPOUND ENGINE AT HAMPTON. HIGH-PRESSURE CYLINDER.

Portion of stroke.	Pressure in pounds absolute.		Actual volume. Cubic feet.	$\theta\phi$ volume (for 1 lb. steam). Cubic feet.
	Forward stroke.	Backward stroke.		
0	99·2	80·1	0·86	0·32
·05	98·3	32·0	2·78	1·05
·1	96·6	22·8	4·70	1·78
·15	93·5	22·8	6·62	2·50
·2	89·7	23·0	8·54	3·23
·25	84·5	23·4	10·46	3·96
·3	74·8	23·6	12·38	4·69
·35	63·5	23·8	14·30	5·41
·4	57·0	23·8	16·22	6·14
·45	51·0	23·4	18·14	6·87
·5	46·1	22·7	20·06	7·59
·55	41·9	21·9	21·98	8·32
·6	38·6	21·2	23·90	9·05
·65	35·8	20·3	25·82	9·77
·7	33·8	19·8	27·75	10·50
·75	31·6	19·4	29·67	11·23
·8	29·8	19·2	31·59	11·96
·85	28·4	19·1	33·51	12·69
·9	27·2	19·4	35·43	13·41
·95	26·1	19·8	37·35	14·14
1·0	24·0	22·5	39·27	14·87

cubic feet, the pressure at that point being 26·12 lb. absolute. The specific volume of dry steam at this pressure is 15·35 cubic feet per pound ; but knowing that the steam is not dry at this point, but possesses a dryness fraction of 0·921, and therefore the volume of 1 lb. of steam of dryness fraction 0·921, and pressure 26·12 lb., will be

$$15\cdot 35 \times 0\cdot 921 = 14\cdot 14 \text{ cubic feet.}$$

TABLE VI.—$\theta\phi$ DIAGRAM FOR COMPOUND ENGINE AT HAMPTON. LOW-PRESSURE CYLINDER.

Portion of stroke.	Pressure in pounds absolute.		Actual volume. Cubic feet.	$\theta\phi$ volume (for 1 lb. steam). Cubic feet.
	Forward stroke.	Backward stroke.		
0	23·3	19·5	1·94	0·76
·05	23·1	4·7	7·19	2·83
·1	22·2	3·5	12·44	4·90
·15	20·8	3·5	17·69	6·96
·2	19·5	3·5	22·95	9·03
·25	18·1	3·5	28·20	11·10
·3	16·7	3·5	33·45	13·17
·35	15·6	2·5	38·70	15·23
·4	14·8	3·6	43·96	17·30
·45	14·0	3·6	49·21	19·37
·5	13·3	3·6	54·46	21·44
·55	12·7	3·6	59·72	23·51
·6	12·0	3·6	64·97	25·57
·65	11·2	3·7	70·22	27·64
·7	10·5	3·7	75·47	29·71
·75	9·8	3·7	80·73	31·78
·8	9·2	3·8	85·98	33·84
·85	8·7	3·8	91·23	35·91
·9	8·3	3·9	96·48	37·98
·95	8·0	4·1	101·74	40·05
1·0	6·5	5·5	106·99	42·11

This will be the $\theta\phi$ volume for that particular point, and is the specific volume of steam at the pressure shown, reduced according to its dryness. The volume occupied by the 0·079 lb. of water will be 0·079 × 0·017 cubic feet, and is quite negligible for all practical purposes. Knowing the $\theta\phi$ volume at any one point, the $\theta\phi$ volumes for the other parts of the stroke are calculated by proportion—*i.e.*, at 10 per cent of the stroke the *actual* volume is 4·70 cubic feet, and the $\theta\phi$ volume will therefore be

$$\frac{4\cdot 70 \times 14\cdot 14}{37\cdot 35} = 1\cdot 78 \text{ cubic feet.}$$

The pressures and volumes for the low-pressure cylinder are similarly dealt with, and the figures are given in Table VI. The $\theta\phi$ diagrams for this trial are given in fig. 11.

It will be seen that in the example quoted no notice has been taken of the increase or reduction in the volume of steam due to the opening and closing of the passage in the main slide valve. This has not been possible in the present case, because the report does not mention when the main valve cut off; but if the position of this had been known, a "kick" would appear on the $\theta\phi$ diagram, somewhat as shown dotted at X, in fig. 11, due to the closing of the passage in the main valve at the end of the forward stroke. The ordinary indicator diagram does not show this.

As a check upon the accuracy with which the $\theta\phi$ diagram has been drawn, its area should be measured by a planimeter, and reduced to heat units according to the scales of temperature and entropy adopted. If the temperature scale be 20 deg. Fah. to 1 in., and entropy 0·1 ϕ = 1 in., then the area of the $\theta\phi$ diagram in square inches, multiplied by 20 × 0·1 = 2·0, will give the work done per stroke expressed in B.T.U. This multiplied by 772, and by the number of strokes made by the engine per minute, and divided by 33,000, should give the same I.H.P. as that calculated from the indicator diagrams for the same cylinder, in the usual way.

12. Complete Entropy Diagram.

Professor Boulvin, of Gand, has investigated and published (see *Engineering*, January 3rd, 1896; and *Revue de Mécanique*, June, 1897) a complete entropy chart for steam, from which, once the curves are correctly drawn, all the remaining points can be obtained geometrically instead of by calculation. The chart is rather complicated, as it includes four separate charts, all drawn from a common centre O (see fig. 12), but it is comprehensive, and saves the continual reference to steam tables.

Let O in the centre of the diagram be the initial point. Plot a temperature scale vertically upwards on the line O W, starting with any convenient minimum temperature, such as 100 deg. Fah. or 32 deg. Fah., and adopting any convenient scale. It will be found advisable to plot the chart on a good-sized sheet of $\frac{1}{10}$ in. squared paper. Draw a scale of entropy O X, horizontally to the right to any convenient scale, the entropy given being above that contained in 1 lb. of water at a temperature equal to the zero adopted in the temperature scale. Volumes are plotted vertically downward on the axis O Y, to a scale to be afterward determined; and pressures (absolute) are measured horizontally to the left, as shown by the line O Z, to any convenient scale.

Starting with the first quadrant (using the word in its trigonometrical sense), the curve O A B, representing the relation between temperature and pressure of dry saturated steam, can be plotted direct from steam tables, using the temperature and pressure scales already decided upon. The entropy curves O C D and E F G, for water and steam respectively, shown in the second quadrant, are drawn as previously described in Chapter II.

To determine the volume scale O Y, take any point on the temperature pressure curve, say A at 250 deg. Fah., and draw a tangent to the curve at A, making an angle a with the vertical axis O W. From C and F in the second quadrant, representing the entropy of 1 lb. of water and

1 lb. dry steam respectively at 250 deg. Fah., draw the vertical ordinates C H and F J on to the axis O X, and from H draw H L parallel to the tangent at A in the first

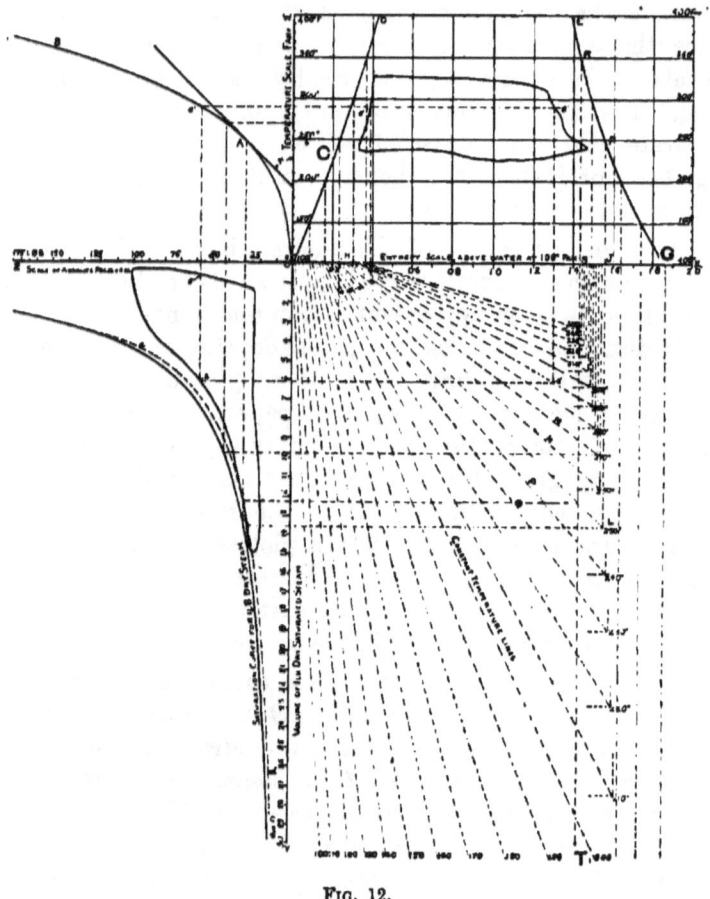

Fig. 12.

quadrant, to meet F J produced in L. Drop similar perpendiculars from other temperatures, such as shown at 200 deg., 300 deg., &c., and draw the inclined lines M N P Q, &c., parallel to tangents to the pressure temperature curve O A B at their respective temperatures.

These lines in the third quadrant represent the increasing volume of 1 lb. steam in being generated at constant temperature, and are therefore called for brevity "constant temperature lines," in contradistinction to the "constant volume lines," drawn on the $\theta \phi$ chart by Captain Sankey's process. They should be drawn to start with a volume of 0·016 cubic feet, and not from the zero of volume scale, as shown; but the volume of 1 lb. water is so small in comparison to the volume of the steam, that it is impossible to show it on the diagram, and is quite negligible. The vertical distances J L, &c., represent to scale the volumes occupied by 1 lb. of dry saturated steam at the temperatures indicated, and furnish the necessary data for constructing the volume scale O Y.

An alternative method of drawing the constant temperature lines, involving a little elementary trigonometry, gives more accurate results than the geometrical method just described. The angle a made by H L to the vertical axis O Y, is equal to the angle a, made by the tangent at A to the vertical axis O W in the first quadrant. But the tangents of all such angles as these are *proportional* to the ratios of $\frac{dp}{dt}$, representing increase in pressure per unit increase of temperature. This ratio of $\frac{dp}{dt}$ for steam is given in "Cotterill on the Steam Engine" (see Appendix, Table IA) for every degree of temperature from 93 deg. Fah. to 432 deg. Fah.; but the ratio given there will only be *equal* to the tan a when the pressure and temperature scales are equal. If the pressure scale be drawn twice as large as the temperature scale (*i.e.*, 1 lb. pressure = 2 deg. Fah.), as it is in fig. 12, the ratio of $\frac{dp}{dt}$ must be doubled, in order to equal the required tangent. Knowing the values of tan a for various temperatures, the lines H L, M, N, P, Q, &c., can easily be drawn with the aid of a protractor and table of natural tangents. Table 7 gives the value of these angles

TABLE VII.—Constant Temperature Lines for Complete Entropy Diagram for Steam.

Temperature Fah., t.	$\dfrac{d\,p}{a\,t}$, from "Cotterill."	Tan a for the particular scales used.	Corresponding value of a.
			Deg. min.
100	0·02835	0·0567	3 15
110	0·03675	0·0735	4 12
120	0·04695	0·0939	5 22
130	0·05945	0·1189	6 47
140	0·07430	0·1486	8 27
150	0·09185	0·1837	10 25
160	0·11335	0·2267	12 46
170	0·1375	0·275	15 23
180	0·1665	0·333	18 25
190	0·2005	0·401	21 51
200	0·2385	0·477	25 30
210	0·2825	0·565	29 28
220	0·3325	0·665	33 37
230	0·391	0·782	38 1
240	0·455	0·910	42 15
250	0·523	1·047	46 19
260	0·603	1·206	50 20
270	0·691	1·383	54 8
280	0·785	1·570	57 30
290	0·892	1·785	60 44
300	1·012	2·024	63 42
310	1·135	2·27	66 14
320	1·275	2·55	68 35
330	1·42	2·84	70 36
340	1·58	3·16	72 26
350	1·75	3·50	74 3

for every 10 deg. Fah., from 100 deg. Fah. to 350 deg. Fah. but must only be used when the pressure scale is twice as open as the temperature scale.

Knowing the volume occupied by 1 lb. of dry saturated steam at a temperature of 250 deg. Fah. (13·53 cubic feet), represented by J L, the volume scale O Y can be drawn, and the saturation curve for 1 lb. of dry steam plotted in the fourth quadrant by the aid of the pressure scale O Z. This completes the construction of the chart as used by Professor Boulvin, which "represents all the transformations of a body by the simultaneous variation of two co-ordinates, which are sufficient to characterise its state, viz., its temperature and its entropy." (See *La Revue de Mécanique*, June, 1897.)

It is instructive to note that the "constant temperature lines" afford a convenient method of drawing the adiabatic expansion curve for steam on the $p.v.$ diagram, without the aid of any calculations whatever. Take 1 lb. of steam, dry, at any particular temperature—say 350 deg. Fah.; its adiabatic expansion can be shown in the fourth quadrant of fig. 12 by the following graphical method : From R, on the steam entropy curve at 350 deg. Fah., draw R S at right angles to O X, and produce it into the third quadrant to T, as shown. Where S T intersects the constant temperature lines will represent the volume occupied by the dry-steam portion of the mixture at various stages of its expansion, and can, therefore, be transferred directly to the pressure-volume curve in the fourth quadrant. By transferring the temperature into the first quadrant, and where it intersects the curve O A B, drawing ordinates into the fourth quadrant gives various points of intersection with the volume lines already drawn, thus giving the adiabatic expansion curve on the $p.v.$ diagram for steam initially dry at 350 deg. Fah. The curve is shown dotted in fig. 12, page 29, and marked a, b, c, d.

13. APPLICATION OF COMPLETE ENTROPY DIAGRAM.

The practical application of the complete entropy diagram will best be explained by taking an actual steam engine test as example. Referring to the trial of a compound engine at Hampton Waterworks, by Professor Beare, already described (see page 23), the mean indicator diagram must first be plotted in the third quadrant to the scales adopted, lengthening or shortening the volume scale as required to correspond with 1 lb. of mixture present in the cylinder. In the trial under consideration, the weight of steam and water passing through the cylinders was 2·4665 lb. per stroke, as found by measuring the water discharged from the air pump. Add to this the weight of steam left in the high-pressure cylinder after compression, from the previous stroke, which is calculated to be 0·1609 lb. per stroke, and the sum, or 2·6274 lb. per stroke, is the weight of steam and water present in the high-pressure cylinder during expansion. This figure also represents a kind of volume factor by which the actual volumes must be divided, in order to obtain the volume of 1 lb. of mixture of similar dryness, for use in plotting on the $\theta\phi$ chart. The actual volume of the high-pressure cylinder at release is 37·35 cubic feet, which, divided by 2·6274, gives 14·215 cubic feet volume at release for an imaginary cylinder containing 1 lb. of mixture of proportionate dryness. Dividing the mean indicator diagram up into any number of equal parts, say 20, of the stroke, the volumes at the other points can be readily transferred to the chart by proportional compasses, or by dividing the actual volume of the cylinder and clearance at any part of the stroke by 2·6274. The diagram being plotted, take any point e on it, and transfer into the first quadrant to meet the curve O A B, fig. 12, and thence into the second quadrant, as shown by dotted lines. Also transfer the point e into the third quadrant, and where it intersects its proper "constant temperature line" transfer into the second quadrant its intersection with the line previously

34   θ φ DIAGRAM FOR CARNOT CYCLE.

drawn at e^1, giving a corresponding point on the entropy diagram. The process of transferring the indicator diagram from the fourth to the second quadrants is one of simple projection, and in reality it takes less time to do it than is occupied in reading how to do it. It should be noted that the difficult curves to be drawn are the fixed ones, whereas the new curves required separately for each experiment are simple ones, and easily drawn with a little practice.

14. θ φ DIAGRAM FOR CARNOT CYCLE.

Having shown how to draw the θ φ diagram for a steam-engine test, it will be interesting to examine the great facilities afforded by it for explaining some of the various

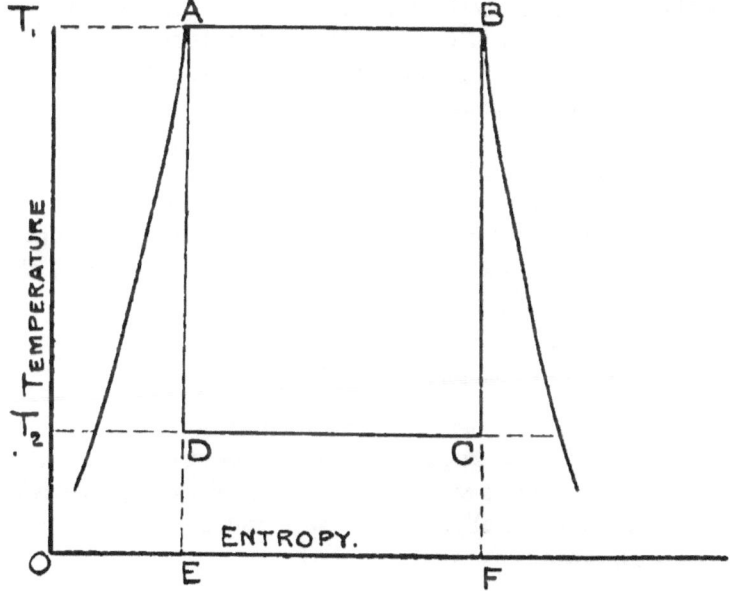

Fig. 13

phenomena that occur in the steam-engine cycle. In the first place, it should be noted that the θ φ diagram proves graphically the efficiency of a perfect reversible cycle. Take the well-known Carnot cycle, shown on the pv or ordinary indicator diagram in fig. 9, page 18, and on the $\theta\phi$ diagram in fig. 13. The work done per stroke is represented in fig. 9 by the area of the figure A B C D, but neither the heat received and rejected, nor its efficiency, are shown on the pv diagram. Transfer the same series of changes to the $\theta\phi$ diagram, shown in fig. 13. The "state point" A represents 1 lb. of water, of temperature τ_1, and entropy O E, evaporated into 1 lb. of dry steam, of the same temperature τ_1, but with an increase of entropy shown by E F. At B, it expands adiabatically to C, its temperature falls to τ_2, but the amount of heat it contains is neither increased nor diminished, and therefore its entropy is the same at C as it is at B; or, the expansion is shown on the $\theta\phi$ diagram by the vertical line B C, representing the fall in temperature from τ_1 to τ_2. At C it is compressed isothermally, at the same temperature τ_2, but giving up a quantity of heat denoted by its reduced entropy from O F to O E. From D to A, adiabatic compression is represented by the vertical line D A, at constant entropy O E, but with an increasing temperature from τ_2 to τ_1. The work done during the cycle is represented in heat units by the area of the rectangle A B C D, fig. 13, and the heat received by the area of the rectangle A B F E; thus the efficiency,

$$\text{or } \frac{\text{work done}}{\text{heat received}} = \frac{ABCD}{ABFE} = \frac{\tau_1 - \tau_2}{\tau_1}.$$

The line O E F corresponds to the absolute zero of temperature; and should be much lower down than indicated on the diagram. The efficiency of this cycle can be proved mathematically, but it is seen much more clearly on the $\theta\phi$ diagram.

15. CONDENSATION DURING ADIABATIC EXPANSION.

The $\theta\phi$ diagram also shows graphically the varying dryness of the steam during the expansion period of the stroke. Neglecting the volume of the water present, which (except in the case of very wet steam) is comparatively nil, the

36 CONDENSATION DURING ADIABATIC EXPANSION.

proportion $\frac{xy}{xx'}$, shown in fig. 11, page 24, represents the dryness fraction of the steam at y, and, similarly, the dryness at any time during expansion can be scaled off the $\theta \phi$ diagram.

It also shows the amount of condensation due to the adiabatic expansion of steam. For instance, 1 lb. of dry steam at 170 lb. absolute pressure, expanded adiabatically to 20 lb.

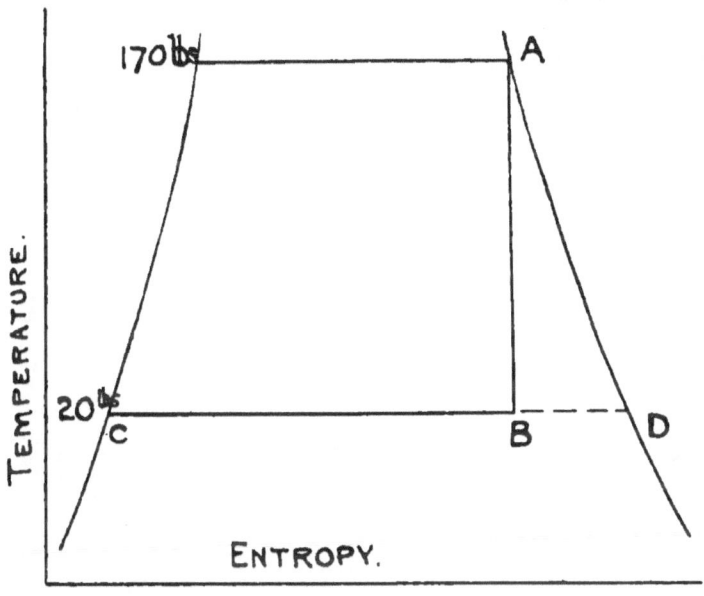

Fig. 14.

absolute, as shown by the line A B in fig. 14, will have a dryness fraction of $\frac{CB}{CD}$, or 0·8797. But the line C B represents the latent heat of 0·8797 lb. steam at 20 lb. pressure, which should occupy a volume of

$$0·8797 \times 19·72 = 17·35 \text{ cubic feet.}$$

According to the law of adiabatic expansion—viz., $p \cdot v^{1·135} =$ constant—the volume occupied by 1 lb. of steam at 170 lb. pressure, initially dry, and expanded to 20 lb., is 17·45 cubic feet. This difference may be due to the volume occupied by, and the heat contained in, the 0·1203 lb. water present, which is not shown by the $\theta\phi$ diagram, but which should be allowed for when dealing with wetter steam.

16. ADIABATIC EXPANSION OF WET STEAM.

If a mixture of steam and water be expanded adiabatically, it is shown on the entropy diagram by dividing the higher temperature entropy into two parts in the ratio of the dryness of the steam at the commencement of expansion. For example, take question No. 41, given in the Science and Art

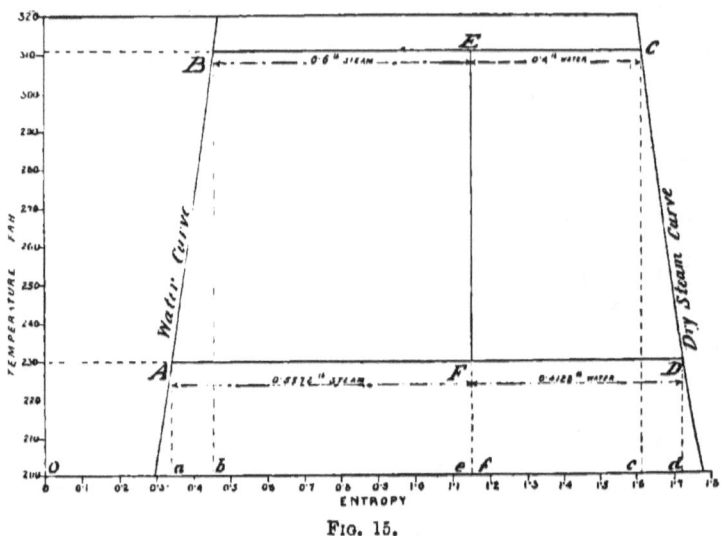

FIG. 15.

Department's Examination in Steam, 1898, where 1 lb. of stuff of 0·6 dryness is expanded adiabatically from 311 deg. Fah. to 230 deg. Fah., and it is required to find the weight of water present at the end of expansion, the entropy of 1 lb. of water being given as 0·339 and 0·451, and that of 1 lb. of dry steam as 1·716 and 1·612 at the lower and higher temperatures respectively. Draw the approximate water and dry steam entropy curves A B and C D from the data given, as shown in fig. 15, and divide B C into two parts representing to scale the dryness of the steam (0·6) at 311 deg. Fah.; that is, make

$$\frac{BE}{BC} = 0·6,$$

and drop the perpendicular EF meeting AD in F at the lower temperature. Then $\frac{AF}{AD}$ will represent the weight of steam, or $\frac{FD}{AD}$ the weight of water present at the end of expansion.

Using italic letters a, b, c, &c., to represent the projection of their corresponding state points on the zero temperature line, the result may be expressed numerically thus:

$$bc = Oc - Ob,$$
$$= 1·612 - 0·451 = 1·161;$$
$$bf = (bc) \times 0·6 = 1·161 \times 0·6 = 0·6966;$$
$$Of = Ob + bf,$$
$$= 0·451 + 0·6966 = 1·1476;$$
$$ad = Od - Oa,$$
$$= 1·716 - 0·339 = 1·377;$$
$$fd = Od - Of,$$
$$= Od - Oe,$$
$$= 1·716 - 1·1476 = 0·5684;$$
$$\frac{fd}{ad} = \frac{0·5684}{1·377} = 0·4128.$$

Answer = 0·4128 lb. water at end of expansion.

It is clearly seen from fig. 15, that if the mixture be very wet to start with (say, consisting of 60 per cent of water and only 40 per cent by weight of steam), adiabatic expansion will produce a dryness, some of the water present becoming re-evaporated at the expense of the sensible heat in the hot water. This would be shown by the ratio $\frac{AF}{AD}$ being greater than $\frac{BE}{BC}$.

CHAPTER IV.—HEAT LOSSES.

17. EFFECT OF STEAM JACKETING.

IN all steam engines a comparatively large proportion of the steam which enters the cylinder is condensed immediately upon its admission, and its work lost to the engine for that period of the stroke. A part of the heat represented by this condensation is returned to the steam towards the end of the expansion period, when its capacity for doing work is considerably diminished; or it may only be returned during the exhaust stroke, when (except in the case of more than one cylinder) it is not only of no use, but impedes the free discharge of the exhaust by increasing the volume of the steam. Various methods have been adopted to reduce this initial condensation to a minimum, the principal of which are steam jacketing, superheating, compounding, and high speed. The various effects of these on the internal working steam of the engine, as shown by the $\theta \phi$ diagram, form a very interesting and instructive study.

The *effect of steam jacketing* is very clearly shown by drawing the $\theta \phi$ diagrams for the same engine, working with and without steam in the jackets. Figs. 16 and 17 are the diagrams for two trials of a triple-expansion engine at the Wapping Pumping Station of the London Hydraulic Power Company, made by Mr. Bryan Donkin in 1892, full particulars of which will be found in the third Report of the Steam Jacket Research Committee of the Institution of Mechanical Engineers, already referred to (see Proceedings of the Institution of Mechanical Engineers, October, 1894, page 536). The cylinders of the engine are 15 in., 22 in., and 36 in. diameter, with a stroke of 2 ft., with pistons coupled direct to water plungers 5 in. diameter. The steam pressure was about 135 lb. absolute, the speed about 56 revolutions per minute, and the total I.H.P. about 175. Comparing the two trials c and d, the former with boiler steam in all the three barrel jackets (the covers are not jacketed), and the latter

EFFECT OF STEAM JACKETING.

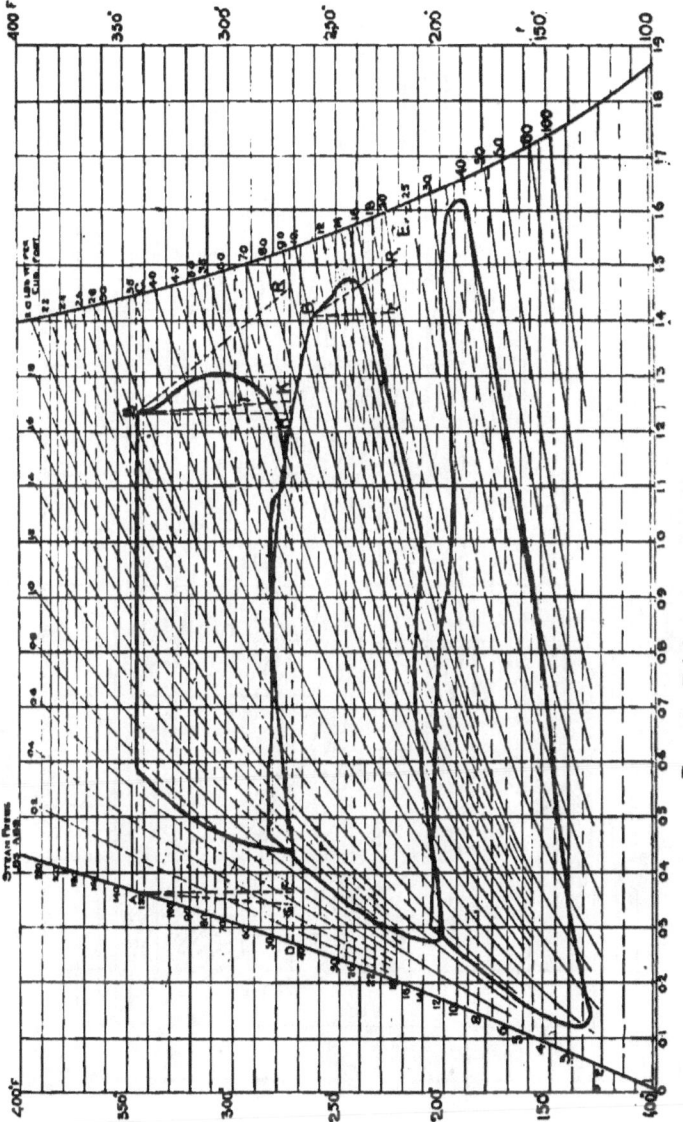

Fig. 16.—Triple-expansion Engine, Jacketed.

EFFECT OF STEAM JACKETING. 41

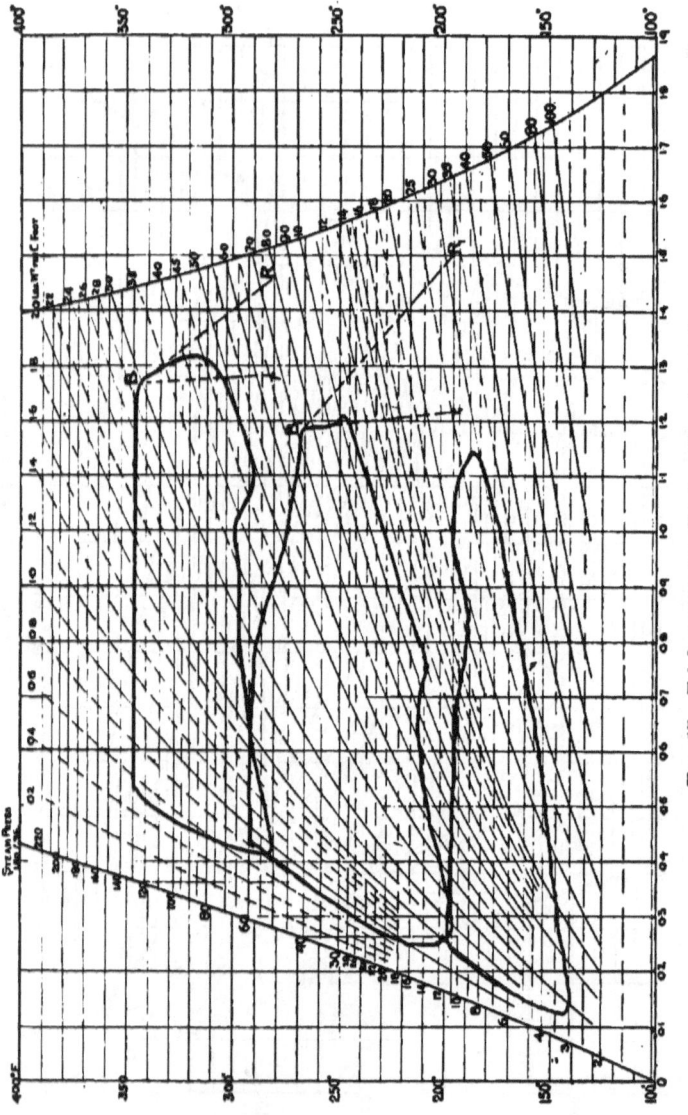

FIG. 17.—Triple-expansion Engine, Non-jacketed.

without steam in the jackets, the consumption of steam per I.H.P. per hour was 15·14 lb. and 17·17 lb. respectively, or 11·8 per cent less in the trial with steam in the jackets.

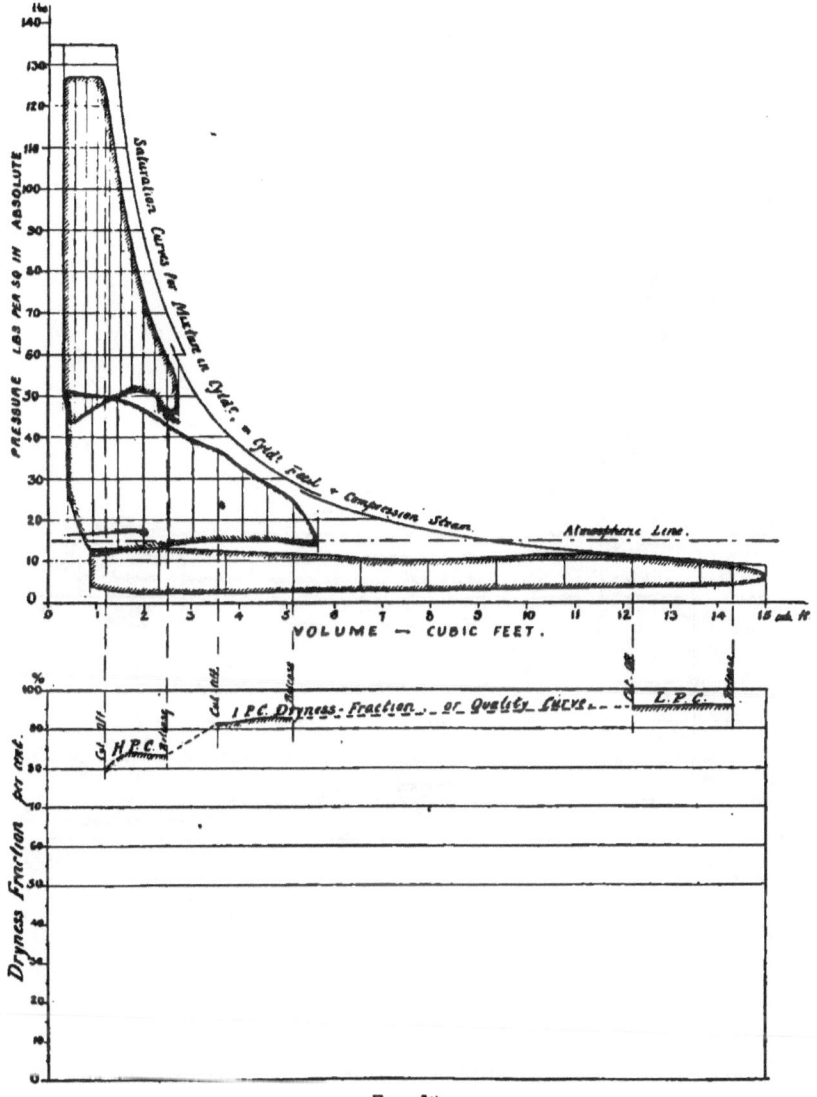

Fig. 18.

EFFECT OF STEAM JACKETING.

The mean indicator diagrams for the two trials are shown in figs. 18 and 19, the former with steam in all jackets and the latter without. The saturation curves for

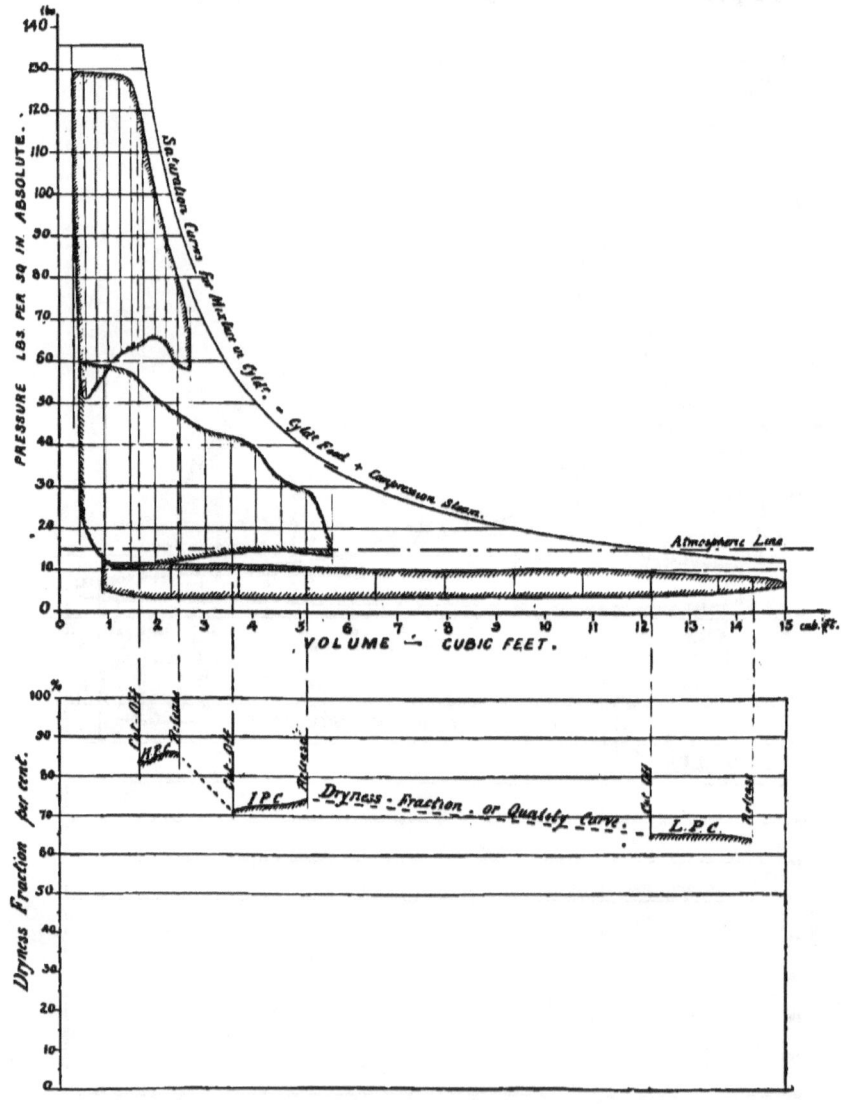

Fig. 19.

the known weights of the mixture of steam and water in each cylinder per stroke have been plotted, in order to show the comparative *work losses* in the two cases. These lines do not form one continuous curve, because the weight of compression or "dead steam" left in the cylinder at the end of each stroke is different for each cylinder; but it is more correct to plot the weight of the mixture than to plot the "live steam" admitted per stroke only, as the comparison of the actual indicator diagram with the former gives the amount of steam present at all points of the stroke. The saturation curves for 1 lb. of dry saturated steam can be plotted if the volumes be taken as the $\theta \phi$ volumes already tabulated to construct the $\theta \phi$ diagram, the table of densities given in the appendix being used for giving the pressures. The curves of dryness fraction shown underneath the mean indicator diagrams have been taken directly from the $\theta \phi$ diagrams, figs. 16 and 17, and show very clearly the gradual condensation and re-evaporation during the progress of the steam from the high-pressure cylinder to the condenser. At certain periods of the stroke both condensation and re-evaporation may be taking place at the same time; in which case, the curves only show the excess of the one action over the other. The $\theta \phi$ diagrams for the trial "c," with steam in the jackets, are shown in fig. 16, and those for trial "d," without steam in jackets, in fig. 17. In comparing the two diagrams, the most noticeable difference is the enormously-increased area of those in fig. 17, especially in the case of the intermediate and low-pressure cylinders. The reason for this is, the water formed during admission to the high-pressure cylinder is gradually re-evaporated by the live steam in the jackets during its passage through the three cylinders, until, when it leaves the low-pressure cylinder for the condenser, it consists of 96 per cent dry steam and 4 per cent of water in the trial with all the jackets on, as compared with 65 per cent dry steam and 35 per cent water in the non-jacketed trial.

EFFECT OF STEAM JACKETING. 45

The comparative areas of the diagrams in figs. 16 and 17 are given in Table VIII., which gives the areas of the $\theta \phi$ diagrams, expressed in B.T.U. per stroke, as compared with the actual I.H.P. for each cylinder, also expressed in B.T.U. per stroke. The mechanical equivalent of heat (J) has been taken as 772 foot-pounds. The volume factor is found by dividing the specific volume of 1 lb. of dry saturated steam, at the pressure shown where the dryness fraction has been calculated, by the actual volume occupied by the steam at that moment. For example, at release in the high-pressure cylinder of trial C, the pressure shown by the indicator diagrams was 59·3 lb. absolute, and the specific volume of dry steam at this pressure is

$$\frac{1}{0·14064} = 7·11 \text{ cubic feet.}$$

Multiply this by 0·855, the known dryness of the steam at that point, and the result is 6·08 cubic feet per pound occupied by the steam. But the volume of the high-pressure cylinder at release is 2·455 cubic feet; therefore the volume factor for the high-pressure cylinder in trial C will be

$$\frac{6·08}{2·455} = 2·476.$$

The volume factor may be looked upon as the reciprocal of the weight of dry steam passing through the cylinder, and the area given in column three of Table VIII. will be the work done in B.T.U. per stroke by 1 lb. weight of steam, which, divided by the volume factor in column 4, gives the work done in B.T.U. per stroke for the known weight of dry steam present in each cylinder. The difference between the figures given in columns 5 and 6 is accounted for by drawing the $\theta \phi$ diagrams in figs. 16 and 17 from a mean indicator diagram which was constructed from nine sets of actual indicator diagrams, taking one set per hour; whereas the figures given in column 6 are calculated from the actual I.H.P. developed in each cylinder, taken from all the indicator diagrams (some forty sets).

TABLE VIII.—Area of $\theta\phi$ Diagrams for Triple-Expansion Engine, with and without Steam in Jackets.

Trial and conditions.	Cylinder.	Area of $\theta\phi$ diagram in B.T.U, measured by planimeter.	Volume factor for each cylinder.	Heat utilised per stroke.	
				Calculated from $\theta\phi$ diagrams. B.T.U.	Calculated from ordinary indicator diagrams.
Trial C. Steam in all jackets.	H.P.C.	51·0	2·47	20·5	20·8
	I.P.C.	68·1	2·865	23·8	23·05
	L.P.C.	58·3	2·82	20·7	20·45
	Total	167·4	....	65·0	64·3
Trial D. No steam in jackets.	H.P.C.	43·5	1·835	23·8	23·6
	I.P.C.	58·45	1·985	29·4	29·4
	L.P.C.	34·52	2·06	16·7	16·3
	Total	136·47	....	69·9	69·3

18. THEORETICAL ENTROPY DIAGRAM FOR SUPERHEATED STEAM.

When saturated steam is removed from the water from which it is generated, and heated beyond the temperature that corresponds to its pressure, it becomes superheated, the additional amount of heat received being

$$Q_1 = Cp \times (\tau_1 - \tau);$$

where Q_1 = heat received as superheat per pound of steam;

Cp = specific heat of superheated steam at constant pressure (usually taken as 0·48);

τ_1 = temperature after superheating;

τ = temperature of saturated steam of the same pressure.

Its entropy is therefore greater than that of corresponding saturated steam, and is equal to

$$\phi_s + 0.48\,(\log_e \tau_1 - \log_e \tau),$$

ϕ_s being the entropy of 1 lb. of dry saturated steam of the same pressure (see Table II., page 13). This increased heat is represented on the entropy diagram, fig. 20, by the extension curve C D, the area under which $CDdc$ (extended to the absolute zero of temperature) is equal to Q_1, or 0·48 $(\tau_1 - \tau)$, where C is taken at temperature τ, and D at τ_1. If expansion be assumed adiabatic—that is, with a non-conducting cylinder and piston—the diagram is completed by the vertical line D E from τ_1 to τ_2, which may intersect

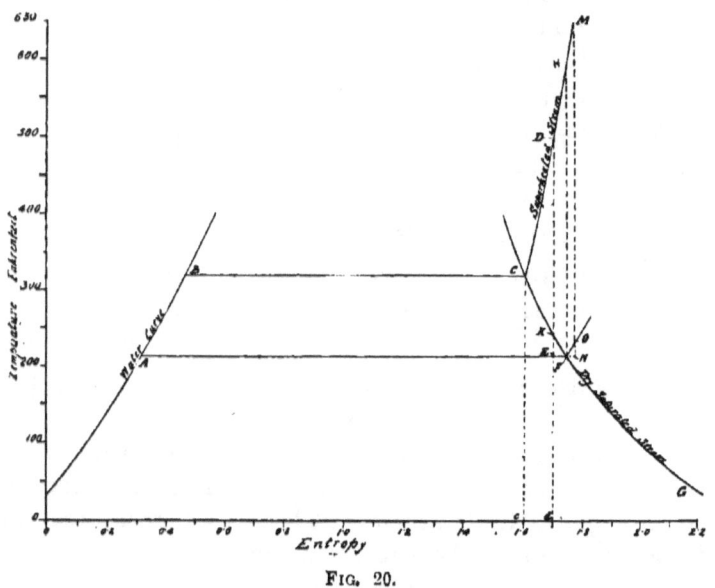

FIG. 20.

the dry saturated steam curve C F G at K, in which case the steam will pass from the superheated to the saturated condition at K, and at the end of expansion be wet steam, with a dryness of $\frac{AE}{AF}$. In order to have dry steam at F, the steam must be superheated up to the point H, the latter being found by drawing the vertical F H to intersect the superheated steam curve C D M. If the superheating be carried beyond this point, as, for instance, to M, the steam remains superheated throughout the whole of the expansion

48 ENTROPY DIAGRAM FOR SUPERHEATED STEAM.

period (assumed adiabatic), and at release it still possesses an amount of superheat represented on the temperature scale by the amount N O, the curve F O being drawn by the same equation as C D M. Table IX. gives the entropy for 1 lb. of superheated steam, starting with dry saturated steam

TABLE IX.—Entropy of Superheated Steam from Dry Saturated Steam at 320 deg. Fah.

Temperature.		Amount of superheat. Deg. Fah.	Increase of entropy. $d\phi$.	Total entropy above water at 32 deg. Fah. ϕ.
Deg. Fah. t.	Absolute. τ.			
320	780	0	0	1·6043
330	790	10	0·0061	1·6104
340	800	20	0·0121	1·6164
350	810	30	0·0181	1·6224
360	820	40	0·0240	1·6283
370	830	50	0·0299	1·6342
380	840	60	0·0356	1·6399
390	850	70	0·0413	1·6456
400	860	80	0·0469	1·6512
410	870	90	0·0524	1·6567
420	880	100	0·0579	1·6622
430	890	110	0·0634	1·6677
440	900	120	0·0687	1·6730
450	910	130	0·0740	1·6783
460	920	140	0·0792	1·6835
470	930	150	0·0844	1·6887
480	940	160	0·0896	1·6939
490	950	170	0·0947	1·6990
500	960	180	0·0997	1·7040
510	970	190	0·1046	1·7089
520	980	200	0·1096	1·7139

at 320 deg. Fah. (corresponding to about 90 lb. absolute pressure), for every 10 deg. up to 520 deg. Fah., or 200 deg. of superheat.

19. EFFECT OF SUPERHEATING.

The effect of superheating the steam, as shown by the $\theta\phi$ diagram, has been very ably treated by Prof. Ripper in his recent paper read before the Institution of Civil Engineers (see Proceedings, vol. cxxviii., January, 1897), and partially reproduced in *The Practical Engineer* (see vol. xvi., page 112).

The main object of using superheated steam is to minimise initial condensation by reducing the exchange of heat between the working steam and the metal walls. When using ordinary saturated steam, which usually contains a small quantity of suspended moisture, condensation during admission must produce a deposit of water upon all the clearance surfaces, which is re-evaporated as the temperature of the steam falls (principally during exhaust), at the expense of the heat stored up in the cylinder walls. The result is, the inner skin of the wall in the clearance surfaces is cooled considerably with every stroke, and condenses a comparatively large proportion of the incoming steam at the next stroke. With superheated steam, the heating up of the walls during admission is done at the expense of the superheat (providing that is sufficiently high); and at cut-off there should still be a little superheat left to evaporate the condensation formed by work being done, so that the steam leaves the cylinder at release just dry, but not superheated. The walls will then be absolutely dry during the exhaust stroke, and consequently require very little heating up at the commencement of the next stroke.

The comparative effect of jacketing and superheating is shown by the four $\theta\phi$ diagrams superposed in fig. 21, which represent four different trials made by Mr. Bryan Donkin on a small vertical engine (see Proceedings of the Institution of Mechanical Engineers, January, 1895, page 132). The

50 EFFECT OF SUPERHEATING.

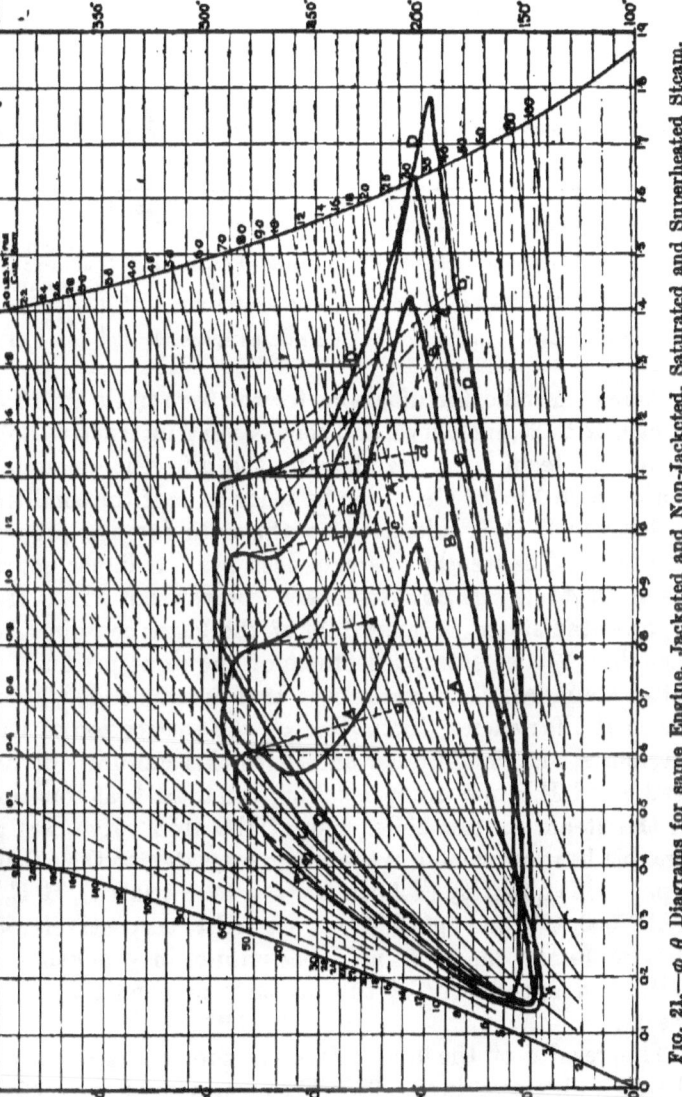

FIG. 21.—$\phi\theta$ Diagrams for same Engine, Jacketed and Non-Jacketed, Saturated and Superheated Steam.

EFFECT OF SUPERHEATING.

conditions and results of the four trials are shown by the following Table X.

TABLE X.

Trial.	Steam in jackets.	Steam in cylinder.	Steam consumpti'n	Remarks.
A	No steam in jackets	Saturated steam	Lb. 45·6	All on the same engine, at same steam pressure, same cut-off, same speed, and same I.H.P.
B	No steam in jackets	Superheated steam	28·4	
C	Saturated steam in jackets	Saturated steam	27·25	
D	Superheated steam in jackets	Superheated steam	20·9	

The mean indicator diagrams for the four trials are shown in figs. 22 to 25, inclusive, and form a very instructive object lesson to steam users, as showing the different amounts of work which 1 lb. of steam can be made to do if the conditions under which it is admitted to the cylinder are properly arranged. The weight of steam used per I.H.P. per hour, as given in table, is not an exact comparison when using superheated steam, because 1 lb. of the latter contains more heat than 1 lb. of saturated steam of the same pressure. A better standard of comparison is that recommended by the Thermal Efficiencies Committee of the Institution of Civil Engineers, viz., the number of British thermal units of heat used per I.H.P. per minute, taking the total heat contained in the steam on the boiler side of the stop valve, less the sensible heat contained in the steam (as water) in the exhaust pipe. This quantity of heat, for the four trials A, B, C, D, is 813, 500, 493, and 368 B.T.U.'s per I.H.P. respectively, and they represent the heat used per minute in producing one I.H.P., independently of the pressure or condition of the steam.

The results of the four trials are shown graphically by their $\theta \phi$ diagrams in fig. 21. The smallness of the diagram in trial A is most noticeable, and shows the enormous loss by condensation during admission which is always produced in

EFFECT OF SUPERHEATING.

non-jacketed cylinders, especially small ones. The partial re-evaporation during expansion, at the expense of the heat

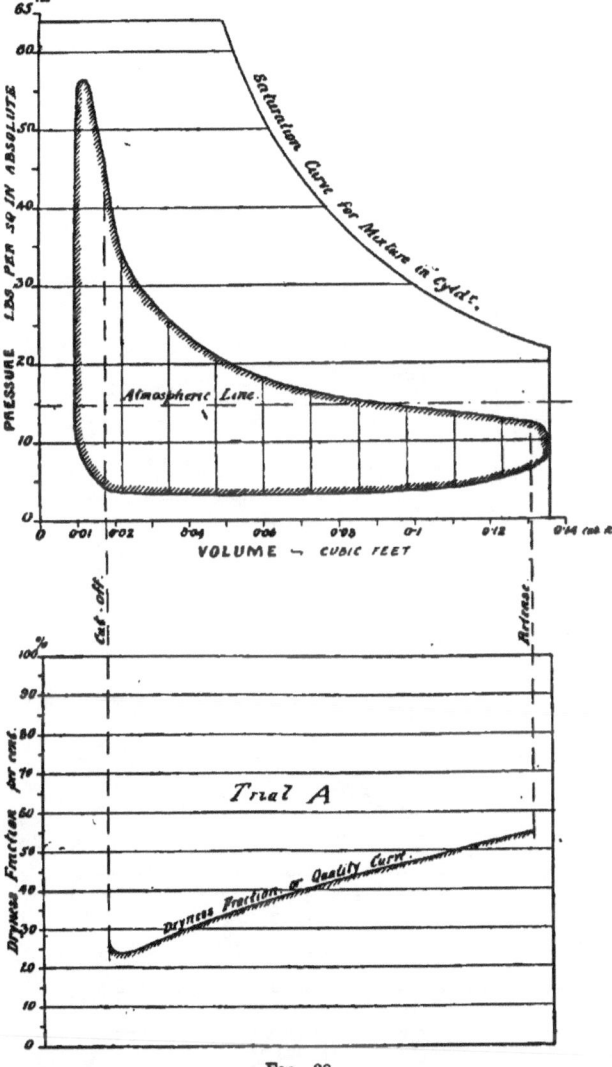

FIG. 22.

in the walls, tends to counterbalance the initial loss; but even at release the weight of steam present in the cylinder

EFFECT OF SUPERHEATING.

is only about one-half what it should have been. The amount of the loss in this particular case was very much increased

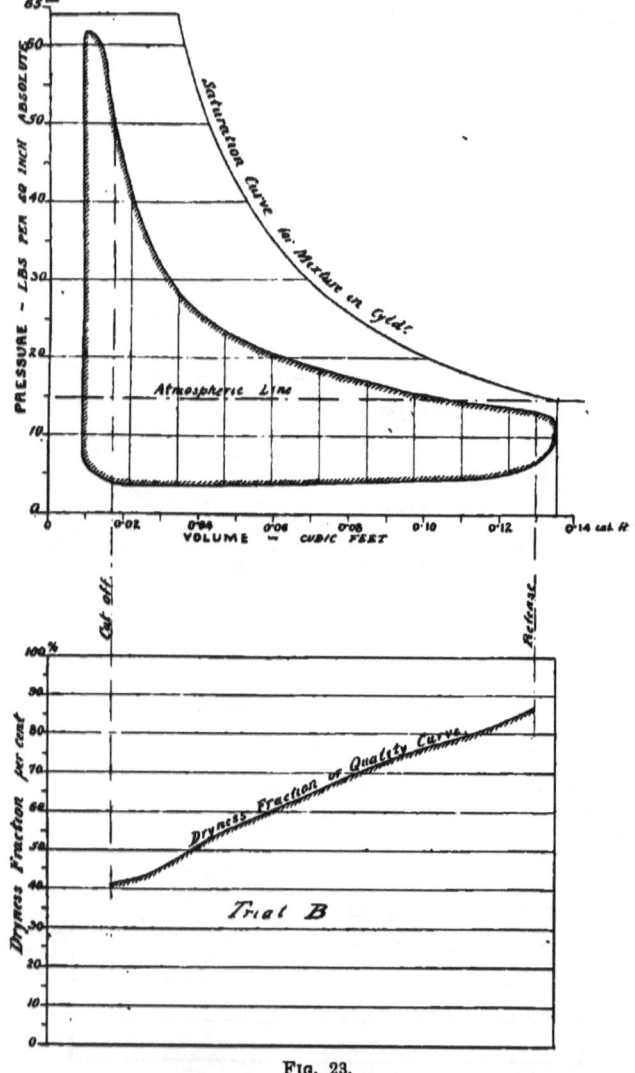

Fig. 23.

by the cut-off taking place very early (at one-sixteenth of the stroke), and the cylinder being so very small (6 in.

diameter, 8 in. stroke); so that the clearance surface, and surface of cylinder exposed to steam during admission, per

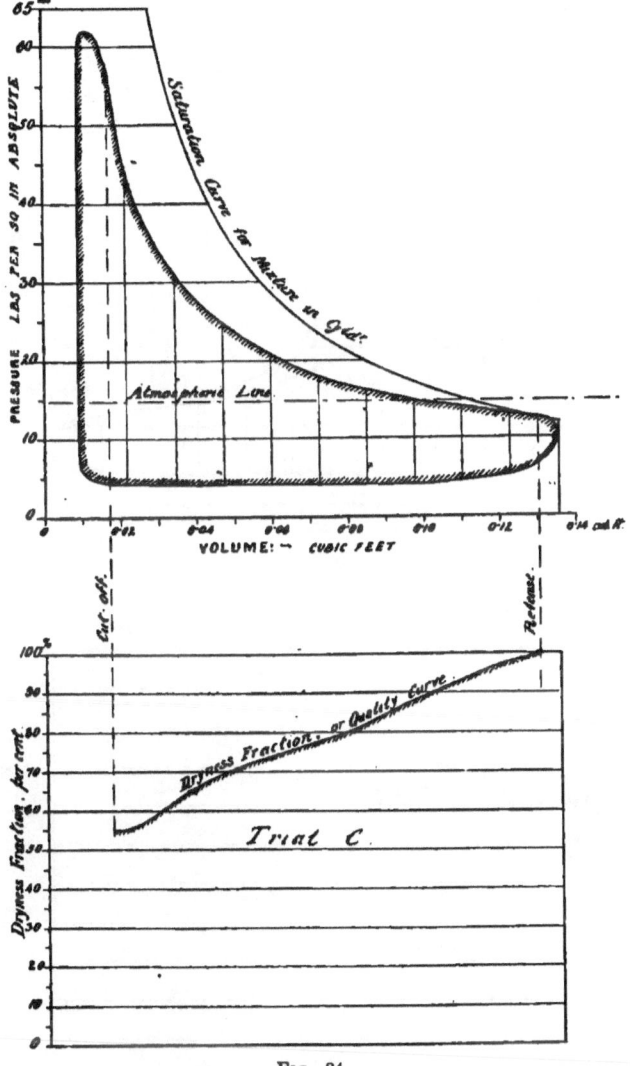

Fig. 24.

pound of steam present, would be very high indeed with a small engine working under these conditions. The conden-

EFFECT OF SUPERHEATING.

sation in a larger engine would be relatively much smaller, as shown by figs. 11 and 17.

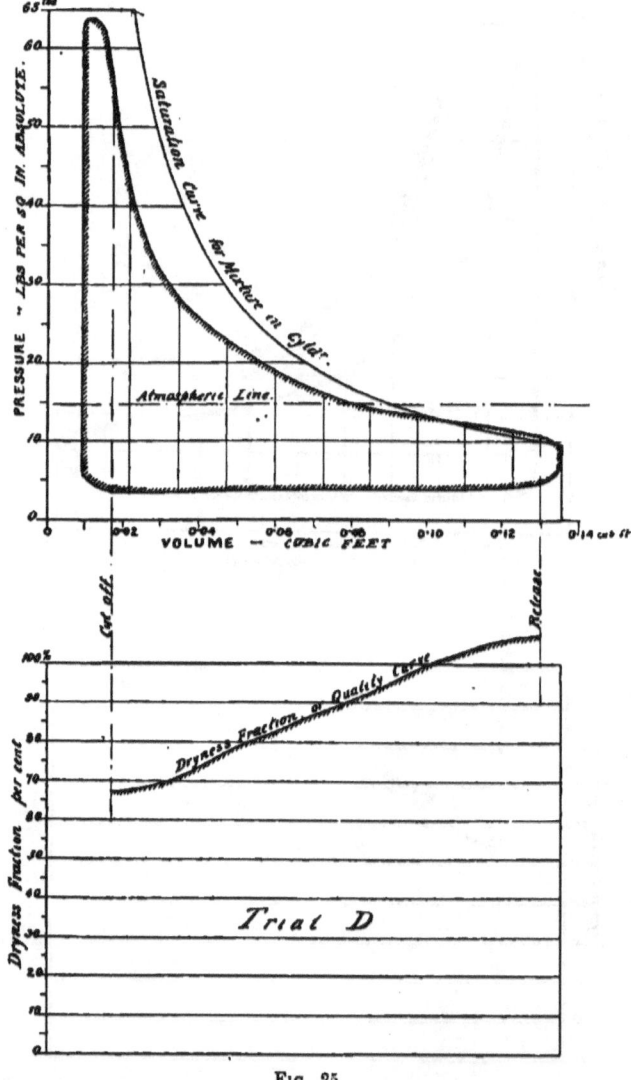

FIG. 25.

In trial B, the diagrams for which are given in fig. 23, with the steam superheated only 34 deg. Fah., and still without steam in the jackets, the reduction effected in the initial condensation is most marked. Instead of there being only 24 per cent of the steam which was admitted present in the cylinder at cut-off, as in trial A, there is now 41 per cent, and the drier walls have produced a larger amount of re-evaporation.

In trial C (see fig. 24), with saturated steam in the cylinder and jackets, there is an initial dryness of 0·53 at cut-off, and the expansion is continued with a gradually-increasing dryness of steam, until at release all the water has been re-evaporated, and it is discharged to the condenser as all dry steam.

In trial D, fig. 25, using steam superheated 59 deg. Fah. in both the cylinder and the jackets, all the water has disappeared by the time the piston has made 0·8 of its stroke, after which the curve crosses the dry-steam boundary, and is slightly superheated. The "toe" of the diagram in trial D has been calculated by assuming that superheated steam follows the law of expansion of a perfect gas, viz., that the product of the pressure and volume varies directly as the absolute temperature, or

$$\frac{p \cdot v}{\tau} = \text{a constant.}$$

20. Effect of Speed.

The effect of increased speed of rotation on the $\theta \phi$ diagram is not very marked. Fig. 28 shows the diagram for two trials on the same engine, at the same steam pressure, cut-off, and general conditions, but in the first trial (shown by the full line in fig. 28) the speed was 216 revolutions per minute, and in the second (shown dotted) at 115 revolutions, or about one-half. With the increased speed there is a little less initial condensation, but the re-evaporation during expansion is greater in the half-speed trial, probably because the time taken by the engine to complete any given portion of its cycle is longer in the latter case, and allows the jackets to make their influence more effective. These

EFFECT OF SPEED.

diagrams are taken from Mr. Bryan Donkin's experiment, Nos. 121 and 122 (see Proceedings of the Institution of

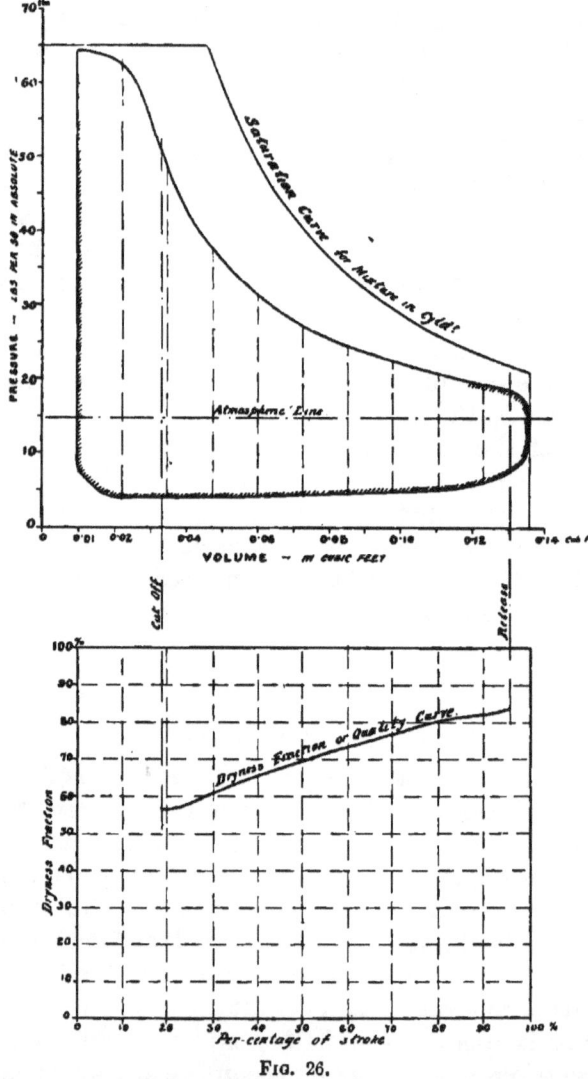

Fig. 26.

Mechanical Engineers, January, 1895). The mean indicator diagrams and quality curves for the two trials are shown in figs. 26 and 27; the former at full speed, and the latter at half speed.

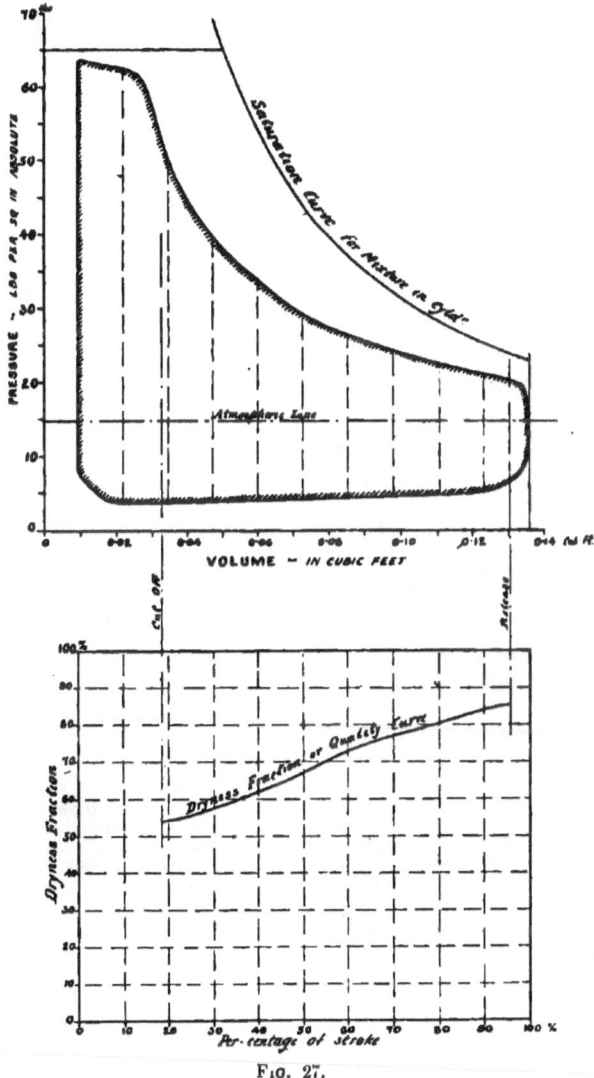

Fig. 27.

The trials of the late Mr. P. W. Willans on his central-valve engine, recorded in the Proceedings of the Institution

EFFECT OF SPEED. 59

of Civil Engineers (vol. xciii., 1888, and vol. cxiv., 1893), prove that the percentage of steam not accounted for by the indicator diagram at cut-off varies inversely as the square root of the number of revolutions per minute. In fact, the amount of initial condensation can be approximately calculated by the following formula:—

$$\frac{W}{I} = c \frac{\log_e r}{D \times \sqrt{N}}$$

where W = the weight of steam per hour *not* accounted for by the indicator at cut-off;
I = the weight of steam per hour accounted for by the indicator at cut-off;
r = number of expansions;
D = diameter of cylinder in feet;
N = revolutions of engine per minute;
c = a numerical coefficient, depending on the design of cylinders and conditions of working.

For ordinary slide-valve engines, with an average amount of clearance surface, c can be taken at about 3 or 4 for jacketed cylinders and 6 for non-jacketed cylinders.

The value of the coefficient c for the two trials just quoted (see figs. 26 and 27) is shown by the accompanying Table XI. to be 2·07 at full speed, and 2·11 at half speed.

TABLE XI.—CONDENSATION AT TWO SPEEDS: OTHER CONDITIONS REMAINING CONSTANT.

Revs. per minute.	Indicated steam at cut-off I. Pounds per hour.	Actual steam used. from experiment. A. Pounds per hour.	Steam condensed during admission.		Condensation coefficient c
			W. Pounds per hour.	Ratio $\frac{W}{I}$	
216·3	130·1	179·4	49·3	0·379	2·07
114·9	68·0	104·1	36·1	0·531	2·11

21. COMPOUNDING.

The use of high-pressure steam, requiring a greater number of expansions for economical working, has necessitated the

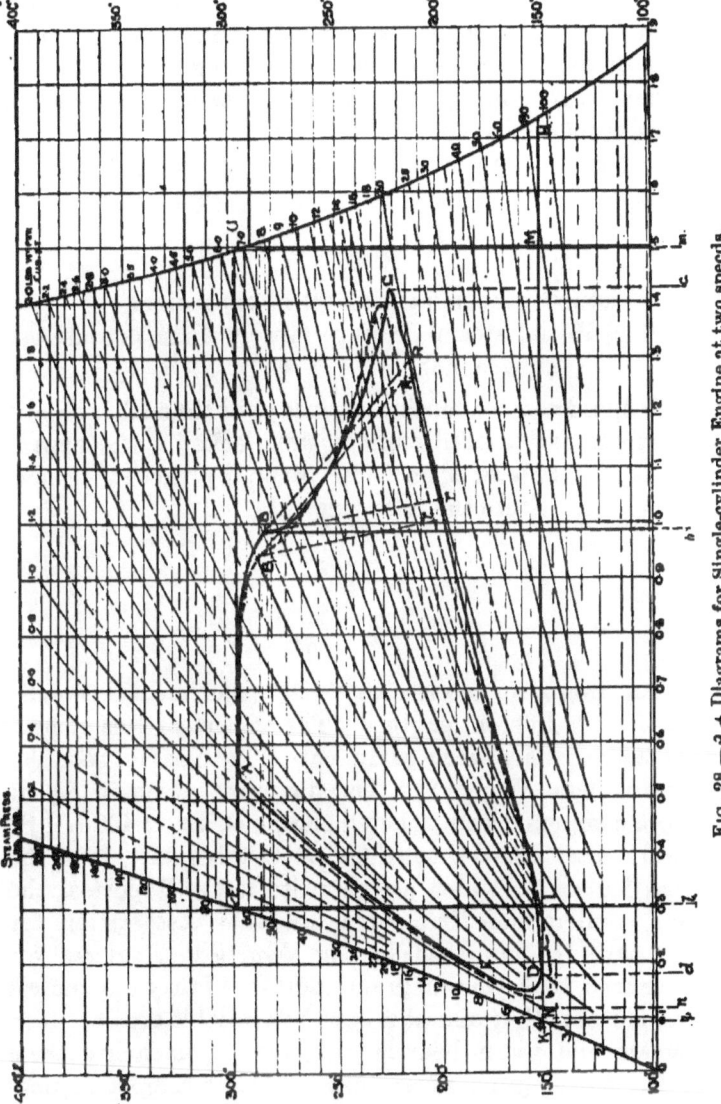

Fig. 28.—$\vartheta \phi$ Diagrams for Single-cylinder Engine at two speeds.

COMPOUNDING.

introduction of multi-cylinder engines, in order that the range of temperature in each cylinder shall not be too large. The question therefore arises, What is the most economical number of expansions—*i.e.*, what is the point beyond which the extra work gained per pound of steam by expansive working is more than neutralised by the loss due to initial condensation? The late Mr. Willans' trials seemed to point to the law—

$$r = \frac{p + 10}{25};$$

where r = most economical number of expansions;
p = initial steam pressure (absolute) in pounds per square inch,

for high-speed non-condensing engines; but the author thinks that with moderate-sized engines, well jacketed, the expansions may be greater than those given by the above law, and would substitute the following rule,

$$r = \frac{p + 20}{20},$$

or
$$r = 1 + \frac{p}{20},$$

as being more in accordance with experimental results obtained with different engines. The enormous loss of heat which accompanies very early cut-off is clearly seen by comparing figs. 31 and 32. The diagrams given in fig. 31 are for a compound non-jacketed engine of about 400 I.H.P., tested by Mr. Michael Longridge on October 21st, 1896, the full particulars of which are recorded in the Report of the Engine, Boiler, and Employers' Liability Association for 1896; partly re-published in *The Practical Engineer* for October 15th, 1897. The diagrams indicate a fairly economical performance, as they almost fill up the available area between the steam and water boundary curves. Comparing them with the $\theta \phi$ diagram given in fig. 32 for a compound non-jacketed engine of the same type, and indicating about the same power, but which was made much too large for the work required, it will be seen how the automatic expansion controlled by the governor destroyed the efficiency of the engine in the latter case (fig. 32). The mean indicator diagrams and quality curves for the two trials are shown in figs. 29 and 30. The cut-off in the high-pressure cylinder for Mr. Longridge's trial was 0·31 of the stroke;

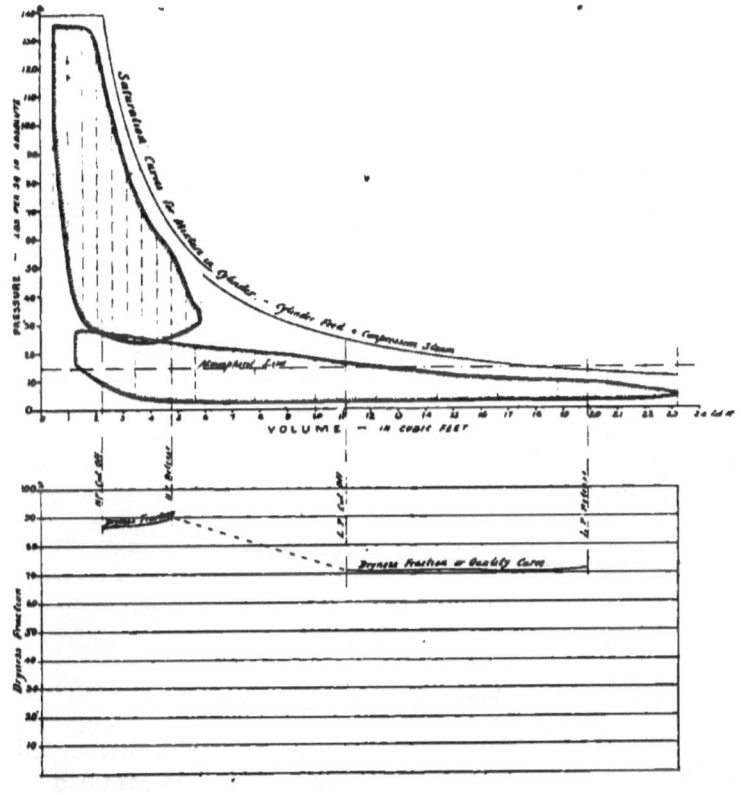

Fig. 29.

and for the diagrams given in fig. 32 the mean cut-off in the high-pressure cylinder was 0·054 of the stroke. The result is, that the steam consumption is increased from 13·9 lb. per I.H.P. per hour in the former, to 27·2 lb. in the latter case. The importance of this waste is realised when one considers that a difference of 13 lb. of steam per I.H.P. per hour for an engine of this size, running night and day, represents a loss of some £15 a week in the coal bill alone.

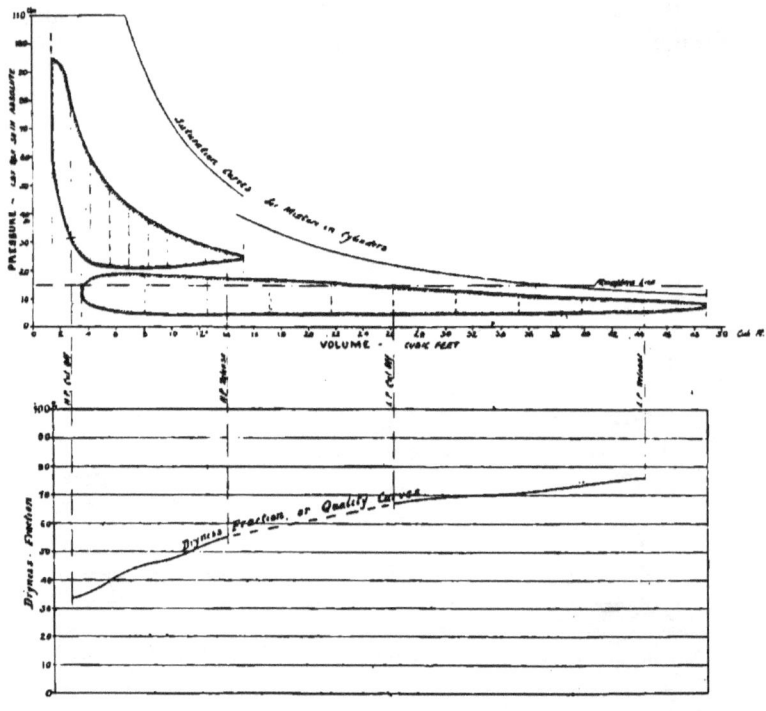

Fig. 30.

22. Initial Condensation.

The weight of steam missing at cut-off may be accounted for in two ways. In the first place, it may be assumed to have entered the cylinder as hot water—*i.e.*, due to priming and condensation in pipes; in which case it would part with its heat during expansion by evaporating a very small portion of the water formed by adiabatic expansion. Assuming a non-conducting cylinder, the expansion curve would then theoretically follow the dotted line Br, shown

64 INITIAL CONDENSATION.

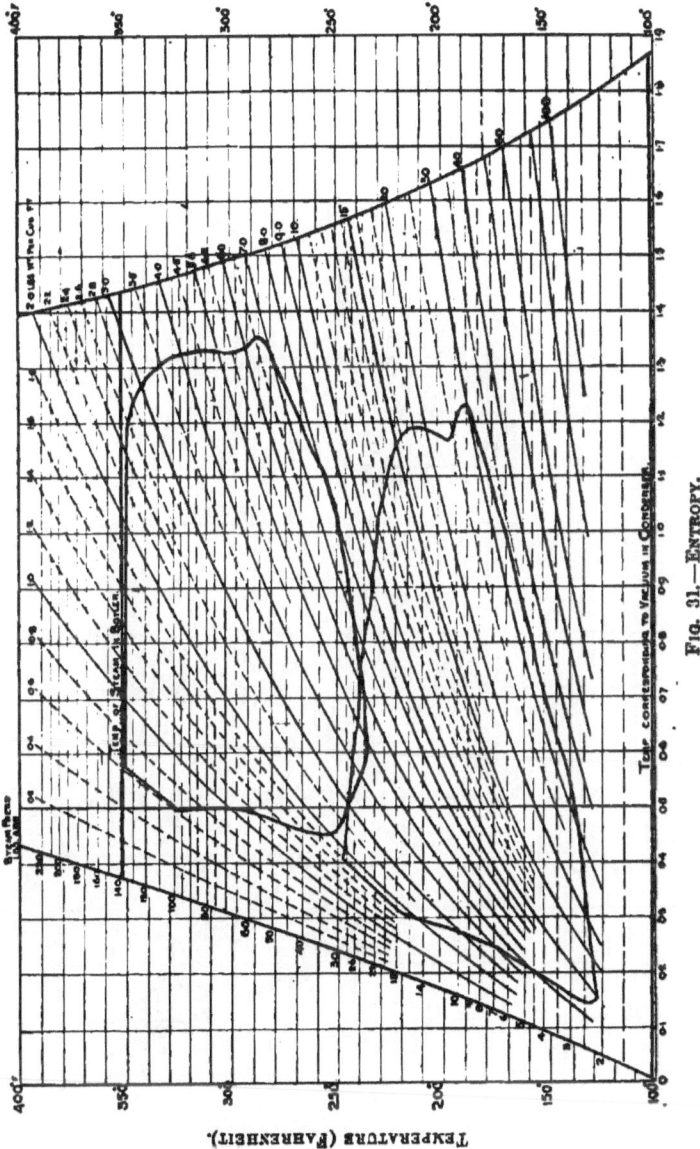

Fig. 31.—Entropy.

INITIAL CONDENSATION.

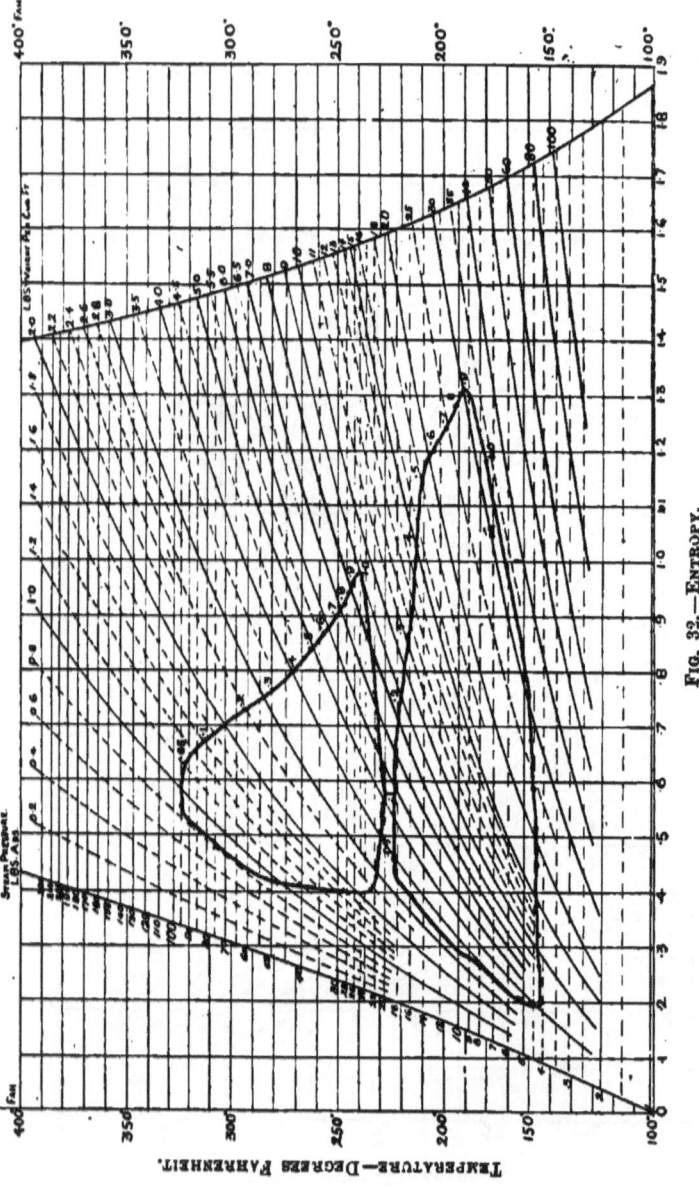

Fig. 32.—Entropy.

on the high-pressure and intermediate cylinder diagrams in figs. 16, 17, &c. This line was called by the late Mr. Willans the "priming water heat-recovery curve" (see discussion by Capt. Sankey of Prof. Beare's paper on "Marine-engine Trials"—Proceedings, Institution of Mechanical Engineers, February, 1894). On the other hand, the quantity of steam missing at cut-off may be assumed to have entered the cylinder as dry steam, and been condensed during admission; in all probability the larger portion of it would be condensed before the stroke had commenced. Assuming the admission valves to be tight, and the cylinder non-conducting, the latent heat of this condensed steam will be stored up in the cylinder walls for use at a later period. If all the heat be returned to the steam during expansion, the $\theta \phi$ diagram from the point of cut-off should follow the dotted line $BR_{\prime}$ shown in figs. 16 and 17. This line is called the "condensation water heat-recovery line." The actual expansion curve of the engine should lie somewhere between these two extreme cases, assuming no external heat to be furnished from a jacket or similar source; in which case the position of the actual curve relatively to Br and BR will show the nature and extent of heat exchange from the metal to the steam after cut-off. Should the cylinder be of small dimensions, or the cut-off very early, this exchange of heat becomes a considerable factor in determining where all the heat of the steam has gone to.

Method of Drawing Br and BR Lines.

Let x represent the fractional dryness of the steam at cut-off, scaled from the $\theta \phi$ diagram; then $(1-x)$ will be the weight of water in pounds present in the cylinder, as the $\theta \phi$ diagram is drawn for 1 lb. of the mixture. But the heat which $(1-x)$ pounds of water will evolve in cooling from a temperature τ_1 to a temperature τ is

$$(1 - x) \cdot s \cdot (\tau_1 - \tau),$$

where s is the mean specific heat of water between τ and τ_1

INITIAL CONDENSATION. 67

Let τ_1 equal the temperature of the steam at cut-off, and τ the minimum temperature of steam in the cylinder. Draw horizontal lines across the $\theta\phi$ chart at these temperatures (see A C and D E, fig. 16), and drop a vertical ordinate from A, meeting D E in F, as shown. Then the triangle D A F, extended to the absolute zero of temperature, represents the heat given to 1 lb. of water in raising it from τ to τ_1; and *vice versâ*, the heat evolved by 1 lb. of water in cooling from τ_1 to τ. Divide D F at G, in the proportion of the dryness fraction at cut-off—*i.e.*, make $\dfrac{DG}{DF}$ equal to $\dfrac{AB}{AC}$; then the triangle G A F, extended to the zero of temperature, will represent the heat given up by $(1-x)$ pounds of water in cooling from τ_1 to τ. This is not mathematically correct, because the curve D A is not quite a straight line; but the error introduced in assuming D A straight is quite negligible, and does not materially affect the length of the line G F. Now turn to the point of cut-off, B, and drop a vertical ordinate from B meeting D E in H, and make the triangle B H K equal to the triangle A G F. The line B H represents adiabatic expansion, assuming a non-conducting cylinder; and the line B K (or B r as it is usually called) will represent the theoretical expansion curve, assuming the $(1-x)$ pounds of water present at B to give up its sensible heat to the steam as the temperature falls; also $\dfrac{HK}{DE}$ will represent, to scale, the fractional weight of water re-evaporated by the heat received from the hot water.

To draw the B R line of condensation water-heat recovery, multiply the weight of water present at cut-off by the latent heat of steam at the same temperature, and divide the product by the mean absolute temperature of the cycle. This will give the amount of increased entropy which would be added to that possessed by the steam present at cut-off, should all the heat of condensation be returned by the walls during expansion.

Using the same notation as before, the increase of entropy will be—

$$\phi_1 - \phi = \frac{(1-x) \cdot L_1}{T_m};$$

where L_1 is the latent heat of 1 lb. steam at τ_1;

T_m is the mean absolute temperature between cut-off and exhaust;

$\phi_1 - \phi$ is the increase in entropy obtained from the re-evaporation.

Referring again to fig. 16, make H R equal to $(\phi_1 - \phi)$, to the scale of entropy used, and join B R, which will represent the theoretical expansion curve, assuming the walls to give back to the steam during expansion all the heat they received from the liquefaction of $(1-x)$ pounds of steam up to cut-off.

The lines B r and B R are shown on the various diagrams, so that the position of the actual expansion line relative to the two theoretical lines can be noted.

23. Measurement of Heat Losses by the Aid of the $\theta \phi$ Diagram.

The most important application of the $\theta \phi$ diagram is, that all the various losses and exchanges of heat can be easily represented graphically, and to a fixed scale of heat units. Referring to fig. 28, and considering only the trial at full speed (shown by the full line), the area of the diagram A B C D E represents in heat units the net work done on the piston per stroke. Draw the maximum and minimum steam temperature lines, F G and K H respectively, and drop the vertical ordinate G M, meeting K H in M; then, if k and m be the ordinates K and M respectively projected on the absolute zero of temperature, the area k K F G M m will represent to scale the total heat supplied to the steam used per stroke, assuming the feed water to be returned to the boiler at the temperature denoted by the line K H. The weight of steam used per stroke is assumed to be 1 lb., for simplicity of calculation; but in reality it is $\dfrac{1}{\text{volume factor}}$

pounds per stroke, as explained previously. The ratio of the two areas

$$\frac{A B C D E}{k K F G M m}$$

will therefore represent the absolute thermal efficiency of the engine under these conditions. If the engine consists of more than one cylinder, the line F G should be drawn at a temperature corresponding to the pressure in the steam pipe near the high-pressure cylinder, and the line K H at the temperature of the hot-well discharge from the condenser. The work done will also be the sum of the areas of the diagrams for each cylinder.

The work done by a perfect heat engine working between the same limits of temperature can be represented by the rectangle L F G M—*i.e.*, the Carnot reversible cycle with adiabatic expansion and compression, and no clearance. The efficiency of the perfect heat engine will therefore be represented on the $\theta \phi$ diagram by the ratio of the two areas

$$\frac{L F G M}{l L F G M m} ;$$

or $\dfrac{\tau - \tau_1}{\tau}$.

The Rankine cycle, which some engineering professors now prefer to use as the standard steam engine of comparison (see report of the Thermal Efficiency Committee of the Institution of Civil Engineers, 1898), includes the heating up of the feed water from the temperature of the exhaust, and is represented in fig. 28 by the area K F G M; the upper limit of temperature, F G, being that corresponding to the pressure of the steam on the boiler side of the stop valve, and the lower limit of temperature, K M, being that of the steam in the exhaust pipe, near the engine. If superheated steam be used, the extra work theoretically obtained should be included, as shown in fig. 20 (page 47). The efficiency of the standard engine of comparison (Rankine cycle) is therefore the ratio of

$$\frac{K F G M}{k K F G M m}.$$

Consequently, the ratio of the actual thermal efficiency to that of the standard heat engine (Rankine cycle) will be represented by

$$\frac{ABCDE}{KFGM}.$$

The ratio of the actual thermal efficiency to that of a Carnot engine working between the same limits of temperature is

$$\frac{ABCDE}{LFGM}.$$

Mr. Willans adopted a fixed minimum pressure for condensing engines, corresponding to a temperature of 110 deg. Fah., in calculating the thermal efficiency of his central-valve engine in his trials, published by the Institution of Civil Engineers (see vol. cxiv., p. 9, Proc.Inst.C.E), and called the thermal efficiency of the engine the ratio of the areas

$$\frac{ABCDE}{KFGM},$$

with FG drawn to correspond to the boiler pressure, and KG drawn at 1·267 lb., corresponding to 110 deg. Fah. This represents the thermal efficiency of the engine compared with a perfect *steam* engine, without clearance, receiving steam at boiler pressure and expanding it adiabatically completely down to the condenser pressure of 1·267 lb., and discharged at that pressure. Mr. Willans' argument was that it was unfair to credit a steam engine with the heat contained by the steam at a temperature (110 deg. Fah.) below which it was useless for power purposes. He therefore compared his engine with a perfect *steam* engine, and obtained a thermal efficiency of some 50 to 60 per cent; whereas the absolute thermal efficiency of steam engines, or the ratio of work done to heat supplied, is only about 12 to 15 per cent.

For the particular trial under consideration, the diagram of which is given in fig. 28, the areas measured by a planimeter gave the following results: Area ABCDE, equal to the net work done per stroke,

$$= 94·85 \text{ B.T.U. for 1 lb. steam.}$$

MEASUREMENT OF HEAT LOSSES.   71

Area $k\,\mathrm{K\,F\,G\,M}\,m$ = heat received per stroke = 1050·2 B.T.U.
Thermal efficiency, or ratio of work done to heat supplied,

$$= \frac{94\cdot85}{1050\cdot2} = 0\cdot0903.$$

Area $\mathrm{L\,F\,G\,M}$ = work done by a perfect heat engine working between same limits of temperature on the Carnot cycle = 171·6 B.T.U. Thermal efficiency of perfect heat engine (Carnot cycle)—

$$\frac{\tau - \tau_1}{\tau} = \frac{171\cdot6}{902\cdot2} = 0\cdot190.$$

Ratio of actual thermal efficiency to thermal efficiency of perfect heat engine (Carnot cycle)

$$= \frac{94\cdot85}{171\cdot6} = 0\cdot553.$$

Work done by perfect steam engine (Willans' method), expanding done to 110 deg. Fah., = 249·6 B.T.U.

Ratio of actual thermal efficiency to the thermal efficiency of a perfect steam engine (Willans' standard)

$$= \frac{94\cdot85}{249\cdot6} = 0\cdot380.$$

The difference is thus seen between the various methods of expressing the thermal efficiency of a steam engine, according to the way in which the term "efficiency" is used and understood.

The areas representing the heat losses in fig. 18 are:

1. Loss due to clearance, equal to area $k\,\mathrm{K\,F\,A\,E\,N}\,n$* (extended to absolute zero of temperature), = 30·0 B.T.U.

2. Loss by heat given to walls during admission = area $b\,\mathrm{B\,A\,G\,M}\,m$; less the heat given by the walls to the steam during expansion, represented by $b\,\mathrm{B\,C}\,c$ = 387·7 - 304·7 = 83·0 B.T.U.

3. Heat carried away by exhaust steam = area $d\,\mathrm{D\,C}\,c$; plus compression $n\,\mathrm{N\,E\,D}\,d$

$$= 801\cdot8 + 40\cdot6 = 842\cdot4\ \mathrm{B.T.U.}$$

---

* The italic letters k, m, n, &c., are the extremities of the ordinates K, M, N, &c., on the line of absolute zero of temperature (− 460 deg. Fah.).

The sum of these quantities (1 + 2 + 3), plus the heat represented by the useful work done (area A B C D E = 94·85 B.T.U.), should be equal to the total heat received by the engine, represented by the area k K F G M m (1,050·2 B.T.U.)

Accounting for all the heat in a balance sheet, according to Hirn's method of treatment, the quantities expressed in B.T.U.'s per stroke for 1 lb. of steam are :—

HEAT BALANCE SHEET.

Heat received.	To —	Heat accounted for.
By steam received. Area k K F G M m. = 1050·2 B.T.U. = 100 per cent.	In useful work .... 1. Loss by clearance...... 2. Loss by walls......... 3. Loss by exhaust steam	94·85 B.T.U. = 9·03 per cent. 30·0 B.T.U. = 2·86 per cent. 83·0 B.T.U. = 7·90 per cent. 842·4 B.T.U. = 80·21 per cent.
	Total	= 1050·25 B.T.U. = 100·0 per cent.

CHAPTER V.—Application to the Gas Engine.

24. General Considerations Applicable to all Permanent Gases.

If 1 lb. of any gas be heated through a small rise of temperature denoted by dt, the amount of heat received can be calculated from its increased energy by the usual formula:

$$dQ = Cv \times dt + A \times P \times dv \quad \ldots \quad (1)$$

where
  dQ = the amount of heat received in B.T.U.;
  Cv = the specific heat of the gas at constant volume;
  $A = \dfrac{1}{J}$, or the reciprocal of Joule's mechanical equivalent of heat;
  P = the pressure of the gas in lbs. per square foot;
  dv = the increase of volume in cubic feet.

The first part of the equation ($Cv \times dt$) represents the additional internal energy of the gas; and the second part ($A \times P \times dv$) that corresponding to the external resistance overcome.

But the combination of Boyle's and Charles's laws proves that the product of pressure and volume of any permanent gas varies directly as the absolute temperature; or

$$P \times V = R \times \tau \quad \ldots \quad (2)$$

where V = volume of the gas in cubic feet;
  R = some numerical constant ($Kp - Kv$);
  τ = absolute temperature.

Substituting the value of P so found in formula (1), we get—

$$dQ = (Cv \times dt) + \left(A \times R \times \tau \times \frac{dv}{V}\right) \quad \ldots \quad (3)$$

and, dividing all through by τ,

$$\frac{dQ}{\tau} = Cv \times \frac{dt}{\tau} + A \cdot \times R \cdot \times \frac{dv}{V} \quad \ldots \quad (4)$$

APPLICATION TO THE GAS ENGINE.

But $\frac{dQ}{\tau}$ is the required small increase of entropy corresponding to the increase of temperature dt.

The external work done by 1 lb. of gas in heating
$$= P \cdot dv = dt\,(Cp - Cv)\,J \text{ (in work units)};$$
therefore $\quad Cp - Cv = A \cdot P \cdot \dfrac{dv}{dt}$,

but also $\quad Cp - Cv = A \cdot R$;

and, substituting this value for $A \cdot R$ in equation (4), we get
$$\frac{dQ}{\tau} \text{ or } d\phi = Cv\,\frac{dt}{\tau} + (Cp - Cv)\,\frac{dv}{V} \quad . \quad . \quad (5)$$

This is the general formula from which the change of entropy of a gas can be calculated under the various conditions of change. For example, during explosion in a gas engine the volume is constant, or very nearly so, and therefore $dv = 0$. Under these circumstances,
$$d\phi = Cv \cdot \frac{dt}{\tau} ;$$
which, integrated, gives
$$d\phi = \int_{\tau_0}^{\tau_1} Cv \cdot \frac{dt}{\tau}$$
$$= Cv \cdot \log_e \frac{\tau_1}{\tau_0} \quad . \quad . \quad . \quad . \quad . \quad (6)$$

On the other hand, if the pressure remains constant, as is sometimes the case just after explosion in a gas engine, the volume and the temperature increase in accordance with Charles's law, viz.,
$$\frac{dv}{V} = \frac{dt}{\tau}, \text{ or } Cv \cdot \frac{dt}{\tau} = Cv \cdot \frac{dv}{V}.$$

In this case, equation (5) becomes—
$$d\phi = Cp \cdot \frac{dv}{V} ;$$
or,
$$Cp \cdot \frac{dt}{\tau} ;$$
which, reduced,
$$= d\phi = Cp \cdot \log_e \frac{\tau_1}{\tau_0} \quad . \quad . \quad . \quad . \quad (7)$$

APPLICATION TO THE GAS ENGINE.

Where both the pressure and the volume of the gas change, as during expansion in a gas engine, the change of entropy must be calculated from the formula representing the law of the particular expansion or compression, viz.,

$$P \cdot V^k = P_1 V_1^k \quad \ldots \ldots \quad (8)$$

But, from equation (2),

$$P \cdot V = R \cdot \tau,$$

and, substituting the value of

$$P = \frac{R \cdot \tau}{V} \text{ in (8)},$$

we get

$$R \cdot \tau \cdot V^{k-1} = P_1 V_1^k \quad \ldots \ldots \quad (9)$$

or, taking the value of $\frac{dv}{V}$ from (3), and substituting in (5), we get

$$d\phi = Cv \cdot \frac{k-\gamma}{k-1} \cdot \frac{dt}{\tau} \cdot \quad \ldots \ldots \quad (10)$$

which, integrated, gives—

$$\phi_1 - \phi_2 = Cv \cdot \frac{k-\gamma}{k-1} \cdot \log_e \frac{\tau_1}{\tau_2},$$

where ϕ_1 = initial entropy;
ϕ_2 = final entropy;
τ_1 = initial temperature;
τ_2 = final temperature.

This is the general formula for finding the change in entropy in all changes represented by the law

$$P \cdot V^k = P_1 V_1^k.$$

Given any expansion or compression curve of an indicator diagram for a gas engine, k is found by measuring any two co-ordinates, thus:

$$P \cdot V^k = P_1 V_1^k;$$

$$\therefore k = \frac{\log P - \log P_1}{\log V_1 - \log V}.$$

In the particular case where $k = \gamma$—that is to say, for adiabatic expansion or compression—the change of entropy becomes *nil*, as $k - \gamma = 0$, and therefore

$$\phi_1 - \phi_2 = Cv \times \frac{k-\gamma}{k-1} \times \log_e \frac{\tau_2}{\tau_1} = 0.$$

The entropy diagram, therefore, for any adiabatic change, is a vertical straight line.

25.—Entropy-Diagram for Theoretical Gas Engine, with Adiabatic Expansion and Compression.

The indicator diagram for a perfect gas engine, working on the Otto or Beau de Rochas cycle, is shown in fig. 33, with pressure as ordinates and volume as abscissæ. In this diagram AB represents the compression of the charge, assumed to be adiabatic—that is, following the law $p \cdot v^\gamma = $ a constant, where γ is the ratio $\frac{Cp}{Cv}$ of the mean specific heats of the gaseous mixture. BC shows the explosion, with instantaneous increase of pressure at constant volume, and CD the expansion period, also assumed to be adiabatic. DA represents the exhaust, and AE the discharge and suction strokes. The entropy diagram for this cycle is shown in fig. 34, and is drawn for 1 lb. of the gaseous mixture, the vertical ordinates representing absolute temperatures (τ) and the base, entropy or ϕ. Starting with the mixture at A, it is compressed from volume Va to Vb adiabatically, without receiving or rejecting any heat. The process is therefore represented in fig. 34 by the vertical line AB, the entropy at B being the same as at A. The pressure at B can be found by the general formulæ—

$$Pa \times Va^\gamma = Pb \times Vb^\gamma;$$

or, taking $\gamma = 1\cdot 4$ (its approximate value),

$$\log Pb = \log Pa + 1\cdot 4 \log Va - 1\cdot 4 \log Vb.$$

The temperature at B is found from the general formulæ—

$$\tau b = \frac{Pb \times Vb}{Kp - Kv},$$

DIAGRAM FOR THEORETICAL GAS ENGINE. 77

where τb = absolute temperature at B;
        Pb = pressure at B in lbs. per square foot;
        Vb = specific volume of the mixture at B;
        Kp = specific heat of the mixture at constant pressure;
        Kv = specific heat of the mixture at constant volume.

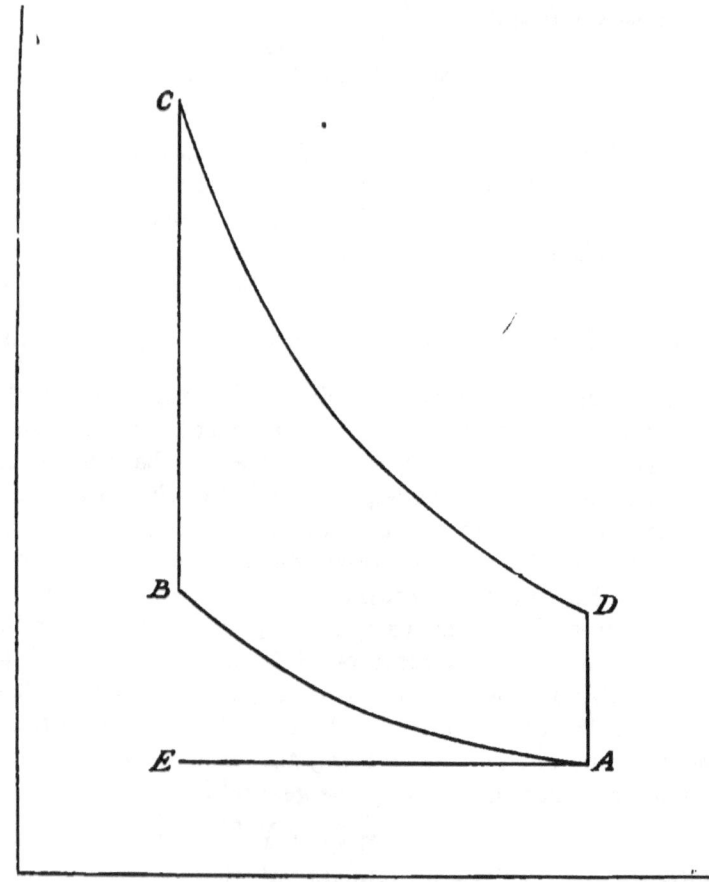

FIG. 33.

[NOTE.—Kp and Kv are in ft.-lb. units, $= J \times Cp$ and $J \times Cv$]

where J = Joule's mechanical equivalent of heat = 772 ft.-lbs.

DIAGRAM FOR THEORETICAL GAS ENGINE.

If the temperature of the mixture at A be known, and the ratio of the compression, the temperature at B can be calculated direct from the general formula for adiabatic compression of a gas—

$$\tau b = \tau a \, (r)^{\gamma - 1}$$

where τb and τa are the absolute temperatures at B and A respectively,

$$r = \text{ratio of compression} = \frac{Va}{Vb};$$

γ = ratio of specific heats.

From B to C the pressure is increased instantaneously before the piston has moved, and therefore the increase of temperature will be directly proportional to the increase of pressure, or

$$\tau c = \frac{\tau b \times Pc}{Pb};$$

Pc and Pb being the absolute pressures in lbs. per square foot at C and B respectively. When the pressure at C is not known, its theoretical temperature can be calculated from the calorific value of the gas, assuming perfect combustion; thus

$$\tau c - \tau b = \frac{H}{R \times Cv}$$

where H = total heat of combustion of 1 lb. of the particular gas used;

R = ratio (by weight) of the gas, air, and diluent to the gas.

In actual practice, this theoretical rise of temperature is, for various reasons, never obtained. An average value of H for London lighting gas is about 19,000 B.T.U.'s per pound. The ratio R varies very considerably with different types of engines and conditions of working, but is about 20 to 30 for lighting gas. The increase of entropy during explosion will be represented in fig. 34 by the logarithmic curve BC, whose equation is

$$d\phi = \int_{\tau b}^{\tau c} Cv \cdot \frac{dt}{\tau};$$

DIAGRAM FOR THEORETICAL GAS ENGINE.

which, integrated, gives

$$\phi c - \phi b = Cv \times \log_e \frac{\tau c}{\tau b};$$

that is, the increase of entropy from B to C is equal to the

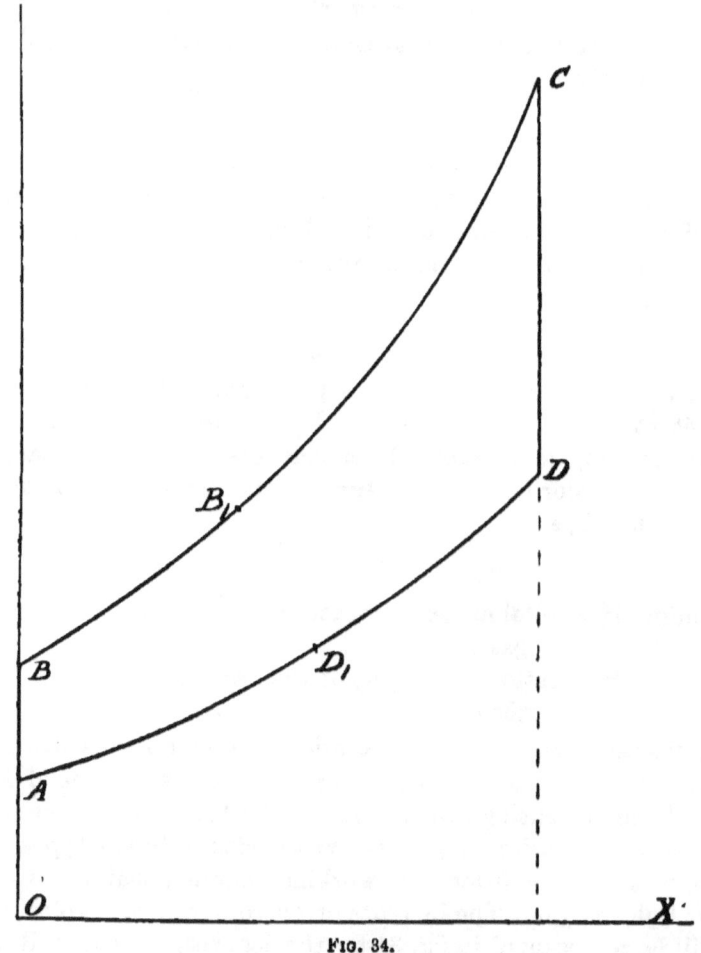

Fig. 34.

specific heat of the mixture at constant volume (about 0·18 to 0·19), multiplied by the hyperbolic logarithm of the ratio of the two temperatures. Knowing τc and $(\phi c - \phi b)$, the

curve B C can be plotted as in fig. 34, all intermediate points such as B_1 being calculated from the formula by substitution, thus—
$$\phi b_1 - \phi b = C v \times \log_e \frac{\tau b_1}{\tau b}.$$
A short geometrical method of constructing such curves as B C and D A will be explained later.

The adiabatic expansion is denoted on the entropy diagram by the vertical line C D, fig. 34, the gas neither receiving nor evolving heat, and the temperature falling from τc to τd; the latter being obtained from the formula,
$$\tau d = \frac{P d \times V d}{K p - K v},$$
and $P d$ being obtained from the usual formula,
$$P d \times V d^\gamma = P c \times V c^\gamma ;$$
or, as $\quad V d = V a$, and $V c = V b$,
$$P d \times V a^\gamma = P c \times V b^\gamma ;$$
or $\quad \log P d = \log P c + \gamma . \log V b - \gamma . \log V a.$

From D to A, or exhaust at constant volume, the entropy diagram again assumes a logarithmic curve D A, fig. 34, the temperature at A falling to its initial point, and its change of entropy being equal to
$$\phi d - \phi a = C v \times \log_e \frac{a}{\tau d}.$$
This change of entropy must always be negative, because $\frac{\tau a}{\tau d}$ is less than unity—that is to say, the *increase* of entropy from D to A is a *minus* quantity or a loss, the gas giving up its heat to the exhaust. The entropy for any intermediate point D_1 on the curve D A can be calculated by substitution:
$$\phi d_1 - \phi d = C v \times \log_e \frac{\tau d}{\tau d_1}.$$
The exhaust and suction strokes A E do not have any effect upon the entropy diagram, as the temperature during those strokes is assumed constant.

The diagram is completed by drawing O X at the absolute zero of temperature, when the work done per cycle will be equal to the area enclosed by A B C D, fig. 34, in heat units ;

the heat received per cycle will be equal to the area $OBCX$, and the thermal efficiency or

$$\frac{\text{work done}}{\text{heat received}} = \text{the ratio of the two areas } \frac{ABCD}{OBCX}.$$

The heat given to the exhaust gases will be equal to the area $OADX$.

It is evident from the entropy diagram that the two quantities $(\phi c - \phi b)$ and $(\phi d - \phi a)$ being equal, and the two curves BC and AD following the same law, the ratio of the two temperatures is a constant quantity, and depends entirely upon the amount of compression—that is,

$$\frac{\tau b}{\tau a} = \frac{\tau c}{\tau d},$$

and the higher this ratio, the higher will be the thermal efficiency.

26.—ENTROPY DIAGRAM FOR ACTUAL GAS-ENGINE TRIAL.

For example, take the trial of a 7 horse power Crossley-Otto engine, made by Professor Capper, at King's College, London, on December 7th, 1892, full particulars of which are given by Mr. Bryan Donkin, in his book on "Gas, Air, and Oil Engines," from which the following data and mean indicator diagram (fig. 35), are taken.

The principal particulars of the trial are:—

Cylinder, 8½ in. diameter by 18 in. stroke.

Revolutions per minute...............	162·5
Explosions ,, ...	71·2
Net I.H.P.	13·32
Cylinder volume	0·591 cubic foot.
Clearance volume.......................	0·2467 ,,
Total volume	0·8377 ,,
Gas used (by meter)	279·75 cb. ft. per hour.
Gas used per explosion, at atmospheric pressure and temperature	0·06544 cubic foot.
Gas used at pressure and temperature in cylinder at A	0·0822 ,,
Air used (from cylinder volume)...	0·7556 cubic foot per explosion.

7T

The mean indicator diagram, fig. 35, furnishes the following pressures and volumes:—

p_a = 13·8 lb. per square inch absolute.
p_b = 67·8 ,, ,,
p_c = 240 ,, ,,
p_d = 240 ,, ,,
p_e = 48·71 ,, ,,
v_a = 0·8377 cubic foot.
v_b = 0·2467 ,,
v_c = 0·2467 ,,
v_d = 0·2617 ,,
v_e = 0·8377 ,,

The point E is taken on the ideal expansion curve, or the actual expansion line continued to the end of the stroke without exhausting. From the above pressures and volumes, the index k in the equation $p_a \times v_a^k = p_b \times v_b^k$ is calculated to be—

For expansion, $k = 1\cdot3707$;
,, compression, $k = 1\cdot3022$.

During compression, the mixture consists of air, gas, and exhaust products, in known proportions, and of known chemical analysis; therefore Kp and Kv for the mixture can be calculated in the same way as for any compound gas. The proportions of air, gas, and diluent present during compression are calculated thus:—

1. Gas, 0·06544 cubic foot per explosion at atmospheric pressure and temperature, with a specific volume of 34·87 cubic feet per pound,

$$= \frac{0\cdot06544}{34\cdot87} = 0\cdot001877 \text{ lb. per explosion.}$$

2. Exhaust products left in cylinder at end of previous discharge stroke = 0·2467 cubic foot at 605 deg. Fah. absolute temperature, and 14·8 lb. per square inch pressure.

From Table XIII., the exhaust gas was of 0·0820 lb. per cubic foot average density.

DIAGRAM FOR AN ACTUAL GAS ENGINE TRIAL. 83

∴ 0·2467 cubic foot weighed 0·2467 × 0·0820,
= 0·02023 lb. at 492 deg. Fah. and 14·7 lb. pressure,

$$= 0.02023 \times \frac{492}{605} \times \frac{14\cdot 8}{14\cdot 7} = 0.01656 \text{ lb.}$$

at 605 deg. Fah. and 14·8 lb. pressure.

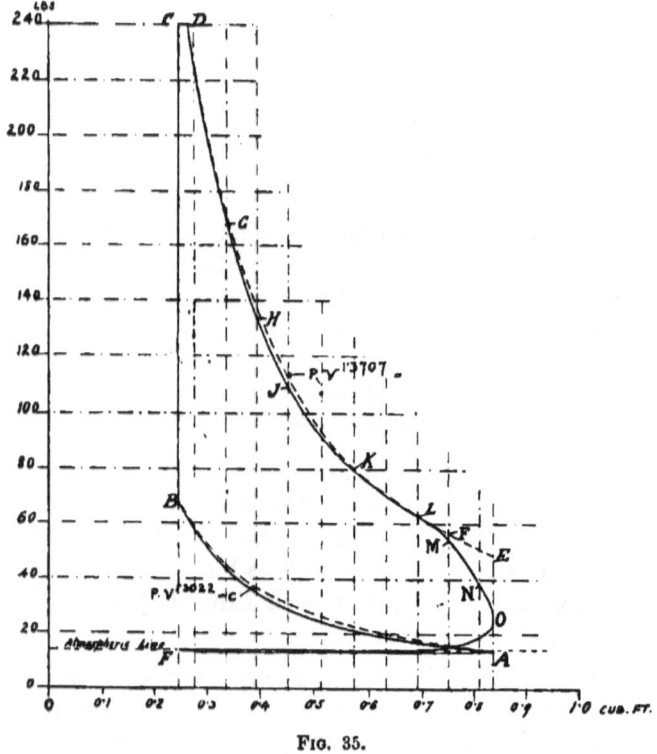

Fig. 35.

3. Air admitted per explosion occupied the total volume of the cylinder, less the volume of the exhaust products and gas. This was

0·8377 − (0·2467 + 0·0822) = 0·5088 cubic feet,

and weighed

$$\frac{0\cdot 5088}{16\cdot 25} = 0.03131 \text{ lb. per explosion.}$$

TABLE XII.—Specific Heats of London Coal Gas, Used in Trial of 7 Horse Power Crossley Engine.

Constituent.	Percentage by volume, per cent.	Weight per cubic foot, lbs.	Weight of in 1 cubic foot of gas, lbs.	Percentage by weight, per cent.	Specific heat at constant volume, Cv.	B.T.U.'s required to raise weight of constituent 1 deg. Fah. B.T.U.	Specific heat at constant pressure, Cp.	B.T.U.'s required to raise weight of constituent 1 deg. Fah. B.T.U.
Marsh gas (CH_4)	31·5	0·0447	0·01408	42·79	0·470	0·2011	0·593	0·2537
Olefines $\{{(C_2H_4) \atop (C_4H_8)}\}$	5·1	0·1174	0·00599	18·21	0·359	0·0654	0·404	0·0736
Hydrogen (H)	51·2	0·00359	0·00286	8·69	2·406	0·2091	3·409	0·2962
Carbon-monoxide (CO)	7·7	0·0783	0·00603	18·33	0·173	0·0317	0·245	0·0449
Nitrogen (N)	3·0	0·0783	0·00235	7·14	0·173	0·0123	0·243	0·0173
Carbon-dioxide $\{{(CO_2) \atop (O)}\}$ and oxygen	1·3	0·123	0·00159	4·84	0·171	0·0083	0·216	0·0104
Totals	..	..	0·03290	100·00	..	0·5279	..	0·6961

DIAGRAM FOR AN ACTUAL GAS ENGINE TRIAL.

TABLE XIII.—SPECIFIC HEATS OF EXHAUST PRODUCTS, FOR TRIAL OF 7 HORSE POWER CROSSLEY ENGINE.

Constituent.	Percentage by volume, per cent.	Weight per cubic foot, lbs.	Weight of in 1 cubic foot of gas, lbs.	Percentage by weight, per cent.	Specific heat at constant volume, C_v.	B.T.U.'s required to raise weight of constituent 1 deg. Fah. B.T.U.	Specific heat at constant pressure, C_p.	B.T.U.'s required to raise weight of constituent 1 deg. Fah. B.T.U.
Carbon-dioxide	6·76	0·123	0·0083	10·17	0·171	0·0174	0·216	0·0220
Oxygen	6·14	0·0895	0·0055	6·70	0·155	0·0104	0·217	0·0145
Nitrogen	87·10	0·0783	0·0682	83·13	0·173	0·1438	0·243	0·2020
Totals	..	..	0·0820	100·00	..	0·1716	..	0·2385

The mixture during compression therefore consisted of
                0·001877 lb. gas ;
                0·01656 lb. exhaust products ;
                0·03131 lb. air.
        Total = 0·049747 lb. per explosion.

The average specific heats of the mixture will be the specific heat of each constituent part multiplied by its relative weight. For the gas, the mean specific heats $C v$ and $C p$ must be calculated from the chemical analysis, as shown in Table XII., from which $C v = 0·5279$, and $C p = 0·6961$; or multiplying by 772, $K v = 407·5$, and $K p = 537·4$. Calculating similarly for the exhaust products, as shown in Table XIII.,

$$C v = 0·1716, \text{ and } C p = 0·2385,$$
or $\quad\quad\quad K v = 132·5$, and $K p = 184·1$.

For air, the figures are $K v = 130·20$, and $K p = 183·55$. The mean specific heat of the mixture can therefore be found—as shown in Table XIV., page 88—to be,
        $K p = 199·09$, and $K v = 141·43$ foot-pounds,
        $n = K p - K v = 57·66$ foot-pounds,
and $C p = 0·25788$,
              $C v = 0·1832$,

$$\gamma = \frac{C p}{C v} = 1·4077.$$

Adopting these values, the temperatures will be—

$$\tau_b = \frac{67·8 \times 144 \times 0·2467}{57·66 \times 0·049747} = 840 \text{ deg. Fah. absolute} ;$$

$$\tau_c = \frac{840° \times 240}{67·8} = 2973 \text{ deg. Fah. absolute} ;$$

$$\tau_d = \frac{2973 \times 0·2617}{0·2467} = 3154 \text{ deg. Fah. absolute} ;$$

$$\tau_e = \frac{48·71 \times 0·8377 \times 144}{57·66 \times 0·049747} = 2048 \text{ deg. Fah. absolute} ;$$

$$\tau_a = \frac{13·8 \times 144 \times 0·8377}{57·66 \times 0·049747} = 580 \text{ deg. Fah. absolute}.$$

DIAGRAM FOR AN ACTUAL GAS ENGINE TRIAL. 87

It will be noticed that the temperature at A found by calculation (580 deg. Fah. absolute) is 25 deg. less than that (605 deg. Fah. absolute) previously assumed for calculating the relative weights of gas, air, and exhaust products in the mixture; but this difference does not affect the result, as all the three constituents will be increased in weight by the same amount, and their relative weights for the corrected temperature will be the same.

To draw the entropy diagram for the trial (see fig. 36), start with the mixture at B as the zero of entropy, when the entropy at C will be found from the formula, as explained previously—

$$\phi_c - \phi_b = C v \times \log_e \frac{\tau_c}{\tau_b}$$

$$= 0.1832 \times \log_e \frac{2973}{840}$$

$$= 0.23158.$$

$$\phi_d - \phi_c = C p \times \log_e \frac{\tau_d}{\tau_c}$$

$$= 0.25788 \times \log_e \frac{3154}{2973} = 0.01524$$

$$\phi_e - \phi_d = C v \times \frac{k - \gamma}{k - 1} \times \log_e \frac{\tau_e}{\tau_d}$$

$$= 0.1832 \times \frac{1.3707 - 1.4077}{0.3707} \times \log_e \frac{2048}{3154}$$

$$= 0.00790.$$

$$\phi_a - \phi_e = C v \log_e \frac{\tau_c}{\tau_e}$$

$$= 0.1832 \times \log_e \frac{580}{2048}$$

$$= -0.23112$$

$$\phi_b - \phi_a = C v \times \frac{k - \gamma}{k - 1} \times \log_e \frac{\tau_b}{\tau_a}$$

$$= 0.1832 \times \frac{1.3022 - 1.4077}{0.3022} \times \log_e \frac{840}{580}$$

$$= -0.02369.$$

88 DIAGRAM FOR AN ACTUAL GAS ENGINE TRIAL.

Balancing up the quantities of entropy, thus—

	+	−
$\phi_c - \phi_b$	0·23158	
$\phi\ \ - \phi_c$	0·01524	
$\phi_e - \phi_d$	0·00790	
$\phi_a - \phi_e$		0·23112
$\phi_b - \phi_a$		0·02369
	0·25472	0·25481

The result shows a numerical error of about 0·04 per cent, which is practically negligible, probably due to slight inaccuracies of calculation.

TABLE XIV.—SPECIFIC HEATS OF MIXTURE, FOR TRIAL OF 7 HORSE POWER CROSSLEY ENGINE.

Constituent.	Weight used per cycle, lbs.	Kv for 1 lb., ft.-lbs.	Kv for weight used, ft.-lbs.	Kp for 1 lb., ft.-lbs.	Kp for weight used, ft.-lbs.
Gas	0·001877	407·5	0·7649	537·4	1·0087
Exhaust products	0·01656	132·5	2·1942	184·1	3·1487
Air	0·03131	130·2	4·0765	183·55	5·7469
Totals	0·049747	..	7·0356	..	9·9043

$K v$ for mixture $= \dfrac{7·0356}{0·049747} = 141·13$ ft.-lbs. per lb.

$K p$ for mixture $= \dfrac{9·9043}{0·049747} = 199·09$ ft.-lbs. per lb.

$n = Kp - Kv = 57·66$ ft.-lbs. per lb.

The $\theta\phi$ diagram for the ideal cycle is shown by A B C D E, in fig. 36, and if the area enclosed be measured by a planimeter, it will be found to equal 171·875 B.T.U. This represents the work which would be done by 1 lb. of gaseous mixture, and has to be corrected by multiplying by 0·049747 to get the work done by the known weight of mixture in the cylinder per explosion. Thus,

$$171\cdot 875 \times 0\cdot 049747 = 8\cdot 55 \text{ B.T.U.}$$

work done per explosion ; or, expressed in work units,

$$8\cdot 55 \times 772 = 6600 \text{ foot-pounds per explosion,}$$

which corresponds within 0·1 per cent with the value given

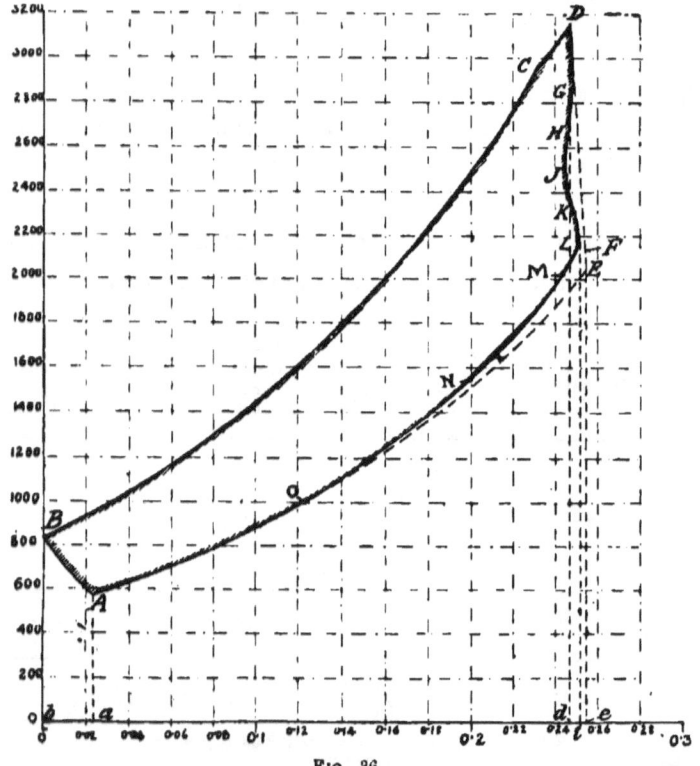

Fig. 36.

by Mr. Donkin (6,594 foot-pounds), as found by measuring the ideal cycle A B C D E of the pressure volume diagram shown in fig. 35.

27.—CORRECTED DIAGRAM FOR GAS ENGINE TRIAL.

This ideal cycle has now to be corrected for the actual diagram, and, starting with the explosion period, the curves

CORRECTED DIAGRAM.

B C and C D, in fig. 36, are correct for the actual diagram, but during the expansion period it will be seen that the actual pressure curve shown by the full line D G H J K L, fig. 35, differs very materially from the ideal expansion curve D E, shown dotted. It will therefore be necessary to take various additional points between D and F upon the actual pressure curve, such as G, H, J, K, and L, and calculate the temperatures and additional entropy at each of

TABLE XV.—Actual Expansion Curve: 7 Horse Power Gas Engine.

Position.	Pressure. Pounds per square inch.	Volume. Cubic feet.	Temperature. Fah. absolute.	Index of expansion. k.	Increase of entropy. $d\phi$.	Total entropy. ϕ.
			deg.			
D	240	0·2617	3154			0·24682
				1·3965	+ 0·000521	
G	170	0·335	2858			0·24734
				1·4668	− 0·00175	
H	134	0·394	2650			0·24559
				1·4798	− 0·00185	
J	109	0·453	2478			0·24374
				1·2995	+ 0·00463	
K	80·5	0·572	2312			0·24837
				1·3526	+ 0·00190	
L	62·5	0·6897	2164			0·25027

these points. For example, at G the pressure is 170 lb. per square inch, and the volume 0·335 cubic feet; the temperature at this point will therefore be

$$\tau_g = \frac{170 \times 144 \times 0·335}{57·66 \times 0·049747} = 2858 \text{ deg. Fah. absolute,}$$

and the entropy at G above that at D will be

$$\phi_g - \phi_d = C v \times \frac{k - \gamma}{k - 1} \times \log_e \frac{\tau_g}{\tau_d}.$$

The value of k, the index of the expansion between D and G, will be

$$k = \frac{\log 240 - \log 170}{\log 0·335 - \log 0·2617} = 1·3965,$$

when $\phi_g - \phi_d$ becomes equal to
$$0.1832 \times \frac{1.3965 - 1.4077}{0.3965} \times \log_e \frac{2858}{3154}$$
$$= 0.000521.$$

The values of k, together with the temperatures, and entropy at the various points G, H, J, K, &c., are given in the accompanying Table XV., from which the actual expansion curve on the $\theta\phi$ diagram, D G H J K, fig. 36, can be plotted and drawn.

For the *exhaust period*, the temperature and entropy can be calculated by the formulæ already given, as follows: At M, just after release, where the actual pressure shown by the indicator diagram is 53·8 lb. per square inch, and the pressure F on the ideal expansion curve at the same volume is 56·79 lb., the two temperatures will be

$$\tau_m = \frac{53.8 \times 144 \times 0.749}{57.66 \times 0.049747} = 2023 \text{ deg. Fah.,}$$

and $\quad \tau_f = \frac{56.79 \times 144 \times 0.749}{57.66 \times 0.049747} = 2135 \text{ deg. Fah.}$

The loss of entropy due to a drop in temperature from 2135 deg. to 2023 deg. at constant volume is equal to

$$d\phi = \phi_f - \phi_m = 0.1832 \times \log_e \frac{2135}{2023} = 0.00991.$$

But this amount of entropy must be taken from that which the mixture would possess at F on the dotted curve. The entropy at F above that at D will be

$$C v \times \frac{k - \gamma}{k - 1} \times \log_e \frac{\tau_f}{\tau_d};$$

$$= 0.1832 \times \frac{1.3707 - 1.4077}{0.3707} \times \log_e \frac{2135}{3154} \text{ deg.}$$

$$= 0.00713$$
add $\quad\quad\quad\quad \phi_c + \phi_d = 0.24682$
then $\quad\quad\quad\quad\quad \phi f = 0.25395$
less $\quad\quad\quad (\phi_f - \phi_m) = 0.00991$
then $\quad\quad\quad\quad\quad \phi_m = 0.24404$

Similar calculations at other points in the exhaust curve of the indicator diagram, such as N and O, are made, as shown in Table XVI., when the actual temperature and entropy at O, coinciding with the theoretical curve E A, finishes the process.

TABLE XVI.—EXHAUST PERIOD: 7 HORSE POWER GAS ENGINE TRIAL.

Position.	Pressure. lbs. abs.	Volume. cub. ft.	Actual Temperature. Fah. abs.	Theoretical Temperature. Fah. abs.	Entropy.
M	53·8	0·749	deg. 2023	deg. 2135	0·25395 − 0·00991 = 0·24404
N	38·0	0·808	1541	1076	0·255 − 0·055 = 0·200
O	24·0	0·8377	1009	2048	0·2557 − 0·134 = 0·1217

28.—CONSTANT VOLUME CURVES.

The two principal curves in the entropy diagram, fig. 36, viz., B C and E A, are seen to follow the general law of a logarithmic curve—

$$\phi_2 - \phi_1 = C \cdot \log_e \frac{\tau_2}{\tau_1},$$

and may be very easily drawn by the following geometrical method. Plot a temperature curve A B C D (fig. 37) to a base of its hyperbolic logarithms, on sectional paper, and between any two temperatures, say at B = 840 deg. Fah. absolute and C = 2,973 deg. Fah. absolute, draw the ordinates B M and C N on to any arbitrary base line X Y. From M draw the inclined line M O P, making an angle a with the base X Y, such that tan a = C, the constant in the above formulæ. Produce M O to intersect the higher temperature ordinate C N at P; then P N will represent to the same scale as the base the change of entropy $\phi_2 - \phi_1$, where τ_2 = 2,973 deg. and τ_1 = 840 deg.

If C be taken = Cv for the trial just considered, viz., 0·1832, then the tan a = 0·1832, or a = 10 deg. 23 min., and P N will be found to scale 0·231. By drawing ordinates from the intersection of the curve with various temperatures between 840 deg. and 2,973 deg., such as 1,000 deg. Fah., 1,200 deg.

CONSTANT VOLUME CURVES.

Fah., &c., the change of entropy from 840 deg. Fah. to these temperatures can be scaled direct from the line M P, being equal to the intersected portions of the ordinates at 1,000 deg. and 1,200 deg., or R S and T U respectively. By adopting this method of graphically finding the intermediate values of entropy for points between B and C, much time and calculation will be saved, and if the curve be drawn accurately and to an open scale, the result will be sufficiently

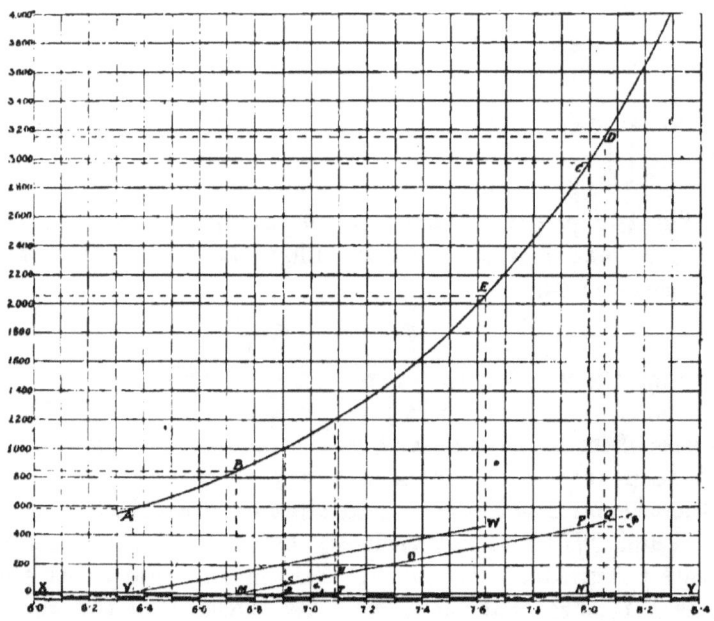

FIG. 37.

correct for all practical purposes. It should be noted that the most laborious part of the process, viz., drawing the logarithmic curve accurately, is only done once for all, and all other temperatures and values of $C p$ and $C v$ can be used on the same chart by drawing various inclined lines M P. The line V W, in fig. 37, is drawn for the period E to A of the 7 horse-power cycle just considered, the entropy at E being 0·23112 above that at A.

94   HEAT LOSSES.

29.—HEAT LOSSES IN 7 H.P. OTTO GAS ENGINE.

In the entropy diagram for the trial of this engine, shown in fig. 36, the heat usefully employed as work is represented by the area enclosed by A B C D J L O; but it does not represent the total heat evolved by the explosion of the gas. Knowing the weight of gas used per explosion (0·001877 lb.), and its calorific value (19,200 B.T.U. per lb.), the total available heat will be

$$0\cdot 001877 \times 19200 = 36\cdot 04 \text{ B.T.U.}$$

This must be represented on the entropy diagram, as in fig. 38, by producing the explosion line B C to P, so that the area $b\,B\,P\,p$ shall be equal to

$$\frac{36\cdot 04}{0\cdot 049747} = 724\cdot 5 \text{ B.T.U. per lb. of mixture.}$$

The theoretical temperature at P, due to complete combustion, can be calculated thus:

Let x = rise in temperature from B,
and C_v = specific heat at constant volume;
then $x \times C_v = 724\cdot 5$,
or $x = 3955$ deg. Fah.,
and temperature at

$$P = 3955 + 840 = 4795 \text{ deg. Fah. absolute.}$$

The several losses can now be estimated direct from the entropy diagram, fig. 38, by measuring the areas they represent, as follows:—

The net work done = A B C D J L O = 8·20 B.T.U. per explosion, or 22·8 per cent of the total available heat. The heat given to the walls during compression, represented by the area $a\,A\,B\,b$, is equal to 0·77 B.T.U. per explosion; that given to the exhaust gases $a\,A\,O\,M\,L\,l$ = 13·63 B.T.U.; and the remainder, or $l\,L\,J\,D\,C\,P\,p$ = 13·44 B.T.U., is transmitted through the cylinder walls. The total heat, therefore, given to the walls is 13·44 + 0·77 = 14·21 B.T.U., and this will be equal to the heat given to the jacket water, plus the radiation of the cylinder and piston. The former of these was measured during the trial, and found to be 14·02 B.T.U. per explosion,

HEAT LOSSES.

thus leaving 0·19 B.T.U. for radiation. It is probable that some of the heat represented by $a\,A\,L\,l$ will have passed through the cylinder wall, and be included in that measured by the jacket water, so that the actual radiation will be

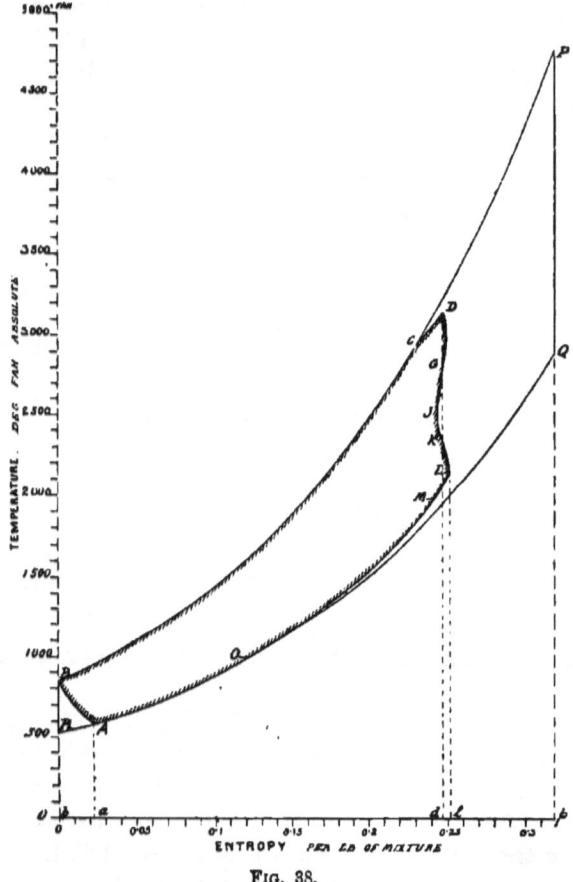

Fig. 38.

considerably in excess of 0·19 B.T.U. The details of the heat balance sheet, as measured from the entropy diagram, fig. 38, are given in Table XVII., and compare very favourably with the values given by Mr. Donkin in his report (Table V., Appendix A, of "Gas, Oil, and Air Engines").

96 HEAT LOSSES.

With an ideal engine, assuming a non-conducting cylinder, complete combustion, and exhaust at constant volume, with adiabatic expansion and compression, the work done per explosion would have been that represented by the area

TABLE XVII.—HEAT BALANCE SHEET FOR 7 H.P. GAS-ENGINE TRIAL.

Area. (See fig. 27.)	Description.	B.T.U. per explosion.	Per cent.
$bBCDd$	Heat received during explosion ...	21·98	
$dDGKLl$	Heat received during expansion ..	0·62	....
$bBCDLl$	Total received shown on diagram ..	22·60	62·7
$bBPp$	Total received, calculated from weight of gas used	36·04	100·0
$LDCPp$	Difference = heat to jacket water..	13·44	37·3
$aAOMLl$	Heat to exhaust gases	13·63	37·8
$aABb$	Heat to walls during compression.	0·77	2·1
	Total heat lost................	27·84	77·2
$ABCDJLO$	Heat in useful work done	8·20	22·8
$bBPp$	Total received................	36·04	100·0

$RBPQ$, instead of $ABCDL$. The maximum work theoretically possible under these conditions is therefore equal to $\frac{T_b - T_a}{T_b}$ of the total heat evolved, amounting in this case to

$$\frac{840 - 580}{840} = 0.3095 \text{ of } 36.04\,;$$

or 11·15 B.T.U. per explosion. The net work actually obtained in the cylinder, being 8·2 B.T.U., was only 73·5 per cent of this. The Carnot cycle of maximum theoretical efficiency, when applied to a gas engine, is therefore not only misleading, but incorrect, because the entropy diagram for a gas engine can never become a rectangle, as in the case of that for the steam engine.

CHAPTER VI.—APPLICATION TO OIL AND AIR ENGINES.

30.—ENTROPY DIAGRAM FOR 20 H.P. DIESEL OIL MOTOR.

IN general, the method of drawing the entropy diagram for an oil-engine test is similar to that employed for the gas engine; the temperature being calculated by the usual formula, $P.V = R.T$, and the entropy from the various equations used in the gas-engine trial, already described (see chapter v., page 75, &c.). As an illustration of a different cycle to the Otto, take that of the Diesel oil motor, described in *The Practical Engineer*, for May 6th, 1898. In this engine the compression stroke is followed by the admission of finely-sprayed oil injected by an air pump, and lasting from 5 to 10 per cent of the explosion stroke, according to the amount of power required. The ignition is effected by heating the air by adiabatic compression to a sufficiently high temperature as to cause the oil to explode. In other respects the cycle is similar to that of an ordinary gas engine. Taking the first trial made by Professor Schröter on a 20 horse power Diesel motor, the following particulars are taken from the *Zeitschrift des Vereines Deutscher Ingénieure* for July 24th, 1897:—

Diameter of motor cylinder ....inches	9·856
Stroke of motor cylinder..................inches	15·725
Capacity of motor cylinder ....cubic feet	0·695
Volume of clearance (assumed 6½ per cent) cubic feet	0·045
Revolutions per minute	171·8
Explosions per minute............................	85·9
Oil used per explosion ....lbs.	0·002122
Air used per explosion ....lbs.	0·039529
Total mixture per explosion ....lbs.	0·041651

Professor Schröter estimates the mean specific heat of the gases at constant pressure at 0·264; and assuming the value of γ to be 1·408, the value of Cv will be 0·1875; or, expressed

8T

in foot-pound units, $Kp = 203.81$, and $Kv = 144.75$, and $R = Kp - Kv = 59.06$ foot-pounds per pound of gas.

From this, knowing the weight of mixture present, and its pressure and specific volume, its temperature can be calculated in the usual way—

$$\tau = \frac{P \times V}{w \times R} ;$$

P = pressure in pounds per square foot (absolute);
V = specific volume of the mixture;
w = weight of mixture per explosion;
R = $Kp - Kv$ for 1 lb. = 59.06 foot-pounds.

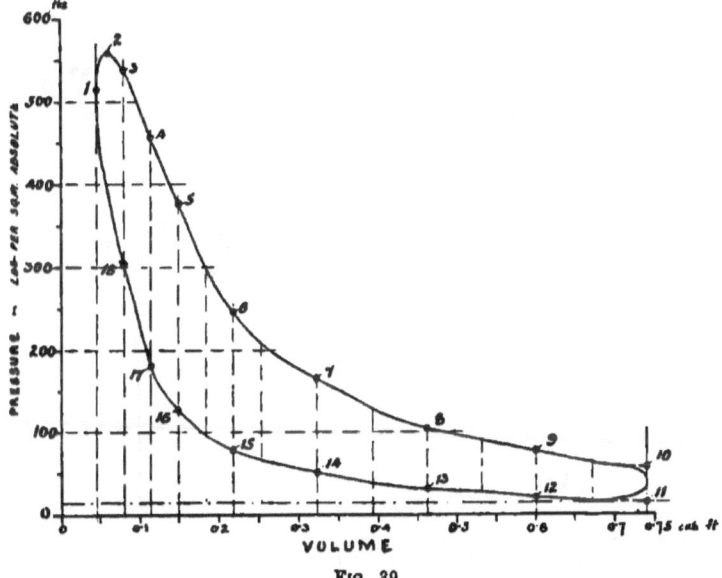

Fig. 39.

Thus, at point 1, at the beginning of the power stroke (see fig. 39), the temperature is—

$$\tau_1 = \frac{515 \times 144 \times 0.045}{0.039529 \times 59.06} = 1430 \text{ deg. Fah.,}$$

and at point 2, where the maximum pressure occurs,

$$\tau_2 = \frac{558 \times 144 \times 0.06}{0.041651 \times 59.06} = 1960 \text{ deg. Fah.,}$$

DIAGRAM FOR DIESEL OIL MOTOR.

assuming the whole of the oil to have been injected during this portion of the stroke.

From 1 to 2 the increase of entropy can best be calculated by the following formula :

$$d\phi = \phi_2 - \phi_1 = Cp \times \log_e \frac{V_2}{V_1} + Cv \times \log_e \frac{P_2}{P_1}$$

$$= \left(0.264 \times \log_e \frac{0.06}{0.045}\right) + \left(0.1875 \times \log_e \frac{558}{515}\right) ;$$

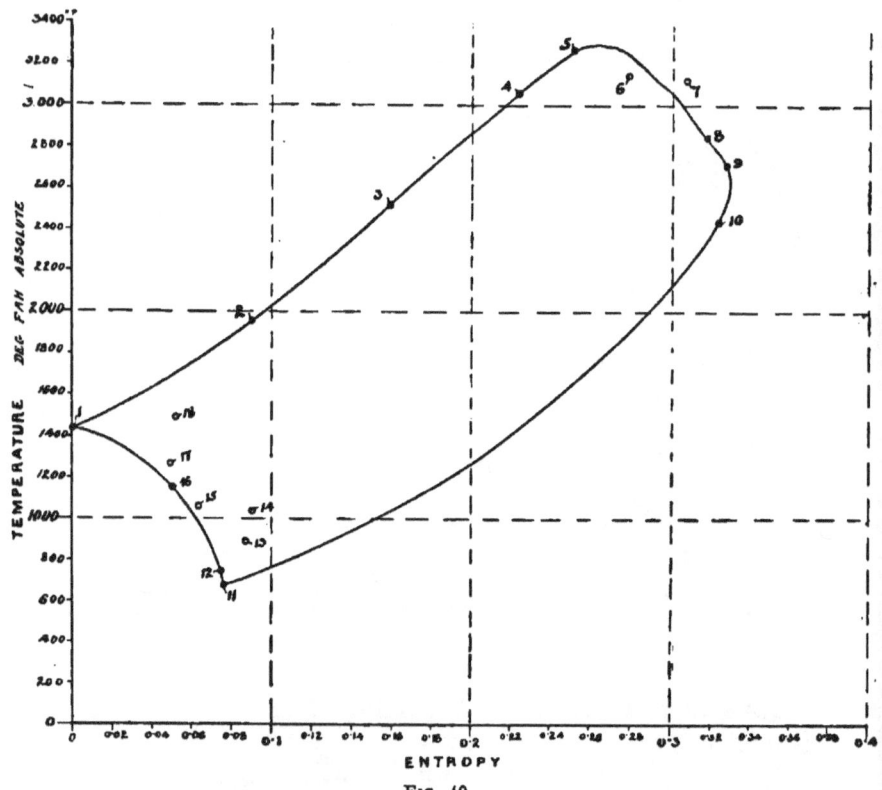

Fig. 40.

that is to say, calculate the difference of entropy due to the increase of volume, as at constant pressure, and add to it the difference of entropy due to the increase of pressure, as at constant volume. After point 2 this second quantity becomes negative, because the ratio of $\frac{P_3}{P_2}$ is less than unity.

TABLE XVIII.—ENTROPY DIAGRAM: 20 H.P. DIESEL MOTOR.

Position.	Pressure Lbs. per square inch.	Volume. Cubic feet.	Absolute temperature. Deg. Fah.	Difference of entropy. Positive.	Difference of entropy. Negative.	Total entropy from position 1.
1	515	0·045	1430	....	....	0
2	558	0·060	1960	0·09098	....	0·09098
3	539	0·0797	2516	0·06863	....	0·15961
4	456	0·1145	3056	0·06413	....	0·22374
5	376	0·1492	3285	0·03380	....	0·25754
6	245	0·2187	3137	0·02062	....	0·27816
7	165	0·323	3120	0·02876	....	0·30692
8	105	0·462	2840	0·00974	....	0·31666
9	77	0·601	2709	0·01128	....	0·32794
10	56	0·740	2426	....	0·00478	0·32316
11	15	0·740	684	....	0·24699	0·07617
12	20	0·601	741	0·00099	....	0·07518
13	31	0·462	883	0·01273	....	0·08791
14	52	0·323	1036	0·00250	....	0·09041
15	73	0·2187	1052	....	0·02686	0·06355
16	125	0·1492	1150	....	0·01250	0·05105
17	180	0·1145	1271	....	0·00160	0·04945
18	305	0·0797	1500	0·00340	....	0·05285
1	515	0·045	1430	....	0·05285	0
Totals	....	....	....	0·34657	0·34657	....

This formula has been used for calculating the entropy throughout the whole of the cycle, because the expansion and compression curves on the $p.v$ diagram do not approach sufficiently to a simple curve of the nature $P_1 V_1^n = P_2 V_2^n$. The *mean* value of n for the expansion period, from 15 per cent of the stroke to the end (points 5 to 10, fig. 39), is 1·1894; but its value varies very much during the expansion. The mean value of n for the compression period (from points 11 to 1) is 1·2629, but that also varies a good deal. The mean indicator diagram from the motor cylinder, shown in fig. 39, is reproduced from the translation of Professor Schröter's trial, as published in the *Engineer* of October 15th, 1897. All the figures for calculating the entropy are given in Table XVIII., and it is satisfactory to note that in this case the positive entropy exactly counterbalances the negative, a good proof of the accuracy of the calculations. The mean effective pressure of the diagram in fig. 39 is 114·5 lb. per square inch, which equals 12,200 foot-pounds of work done per explosion. The area of the entropy diagram in fig. 40 is equivalent to 11,900 foot-pounds of work per explosion, or $2\frac{1}{2}$ per cent less than that shown by the $p.v$ diagram. This difference is probably due to the various assumptions made in calculating the specific heats. These calculations do not include the power absorbed by the air pump, which must be taken into account when calculating the net I..H.P., to obtain the consumption of oil.

31.—APPLICATION TO STIRLING'S HOT-AIR ENGINE.

Air engines may be divided generally into two classes— (*a*) those in which the air is heated at constant volume, and (*b*) those in which it is heated at constant pressure. Of the former class, Stirling's engine is perhaps the most common example; and the cycle of operations for such an engine is shown on the pressure volume diagram in fig. 41, and on the temperature entropy diagram in fig. 42. In this engine the first operation is to admit a quantity of heated air at a temperature τ_1 from the regenerator to the motor cylinder, and expand it isothermally at the higher temperature τ_1 from A to B. The loss of heat due to the work done during expansion is repaired by an external furnace, so as to keep its temperature constant during the stroke. At B the air is passed through the regenerator, where it deposits some of its heat ; its temperature falling from τ_1 to τ_2, and its pressure from B to C at constant volume. At C communication with the regenerator is closed, and the cooled air is compressed isothermally at the lower temperature τ_2, indicated by the curve CD ; and, finally, it is passed through the regenerator again, to take up its deposited heat at constant volume ; its temperature rising from τ_2 to τ_1, and its pressure from D to A. The

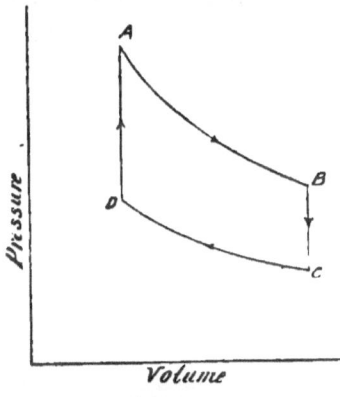

Fig. 41.

thermal changes which take place will be better understood by reference to fig. 42, which shows the entropy diagram for the same cycle lettered similarly to fig. 41. Here the straight lines AB and CD represent the isothermal expansion and compression at τ_1 and τ_2 respectively ; and BC and DA the constant volume curves, whose equation is :—

$$\phi = Cv \times \log_e \frac{\tau_1}{\tau_2}$$

where ϕ = change of entropy between any two temperatures τ_1 and τ_2,

and Cv = specific heat of air at constant volume
       = 0·1686 B.T.U. per lb.

The heat supplied to the air during expansion is represented in fig. 42 by the area $ABba$; and that deposited by

STIRLING'S HOT-AIR ENGINE. 103

the air in the regenerator by $cCBb$. That rejected during compression is shown by the area $dDCc$; and that taken up in the regenerator by the air at the end of the cycle by the area $dDAa$. The net work done during one cycle is represented in heat units by the enclosed area $ABCD$, and the net heat supplied by the furnace (excluding the regenerator, which only acts as a reservoir of heat) being $ABba$, the efficiency of the cycle will be the ratio

$$\frac{ABCD}{ABba}.$$

But the two curves DA and CB are similar—that is, they are both logarithmic curves of the same equation between

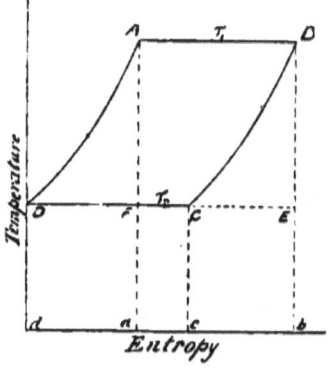

Fig. 42.

the same temperature limits, and therefore the area DAF will be equal to the area CBE. This must be so, because the amount of heat received from the regenerator is equal to the amount deposited in it, radiation being neglected. The area $ABCD$, representing the work done, will therefore be equal to the area of the rectangle $FABE$; and the efficiency becomes equal to

$$\frac{FABE}{ABba} = \frac{\tau_1 - \tau_2}{\tau_1}.$$

Thus it can be geometrically proved without any mathematics that the Stirling hot-air engine has theoretically a maximum possible efficiency, being equal to that of the Carnot reversible cycle.

32.—Application to Ericsson's Hot-Air Engine.

The other class of air engines, viz., those that receive heat at constant pressure, can be similarly treated. For example, take the cycle of Ericsson's engine, shown with pressure and volume as ordinates in fig. 43, and temperature entropy ordinates in fig. 44. In this case, the first stage is admission from A to B of hot air at constant pressure, with its consequent reception of heat, represented in fig. 44 by a A B b. Then expansion isothermally at the higher temperature τ_1 from B to C, followed by the exhaust at constant pressure from c to D, the heat rejected being equal to

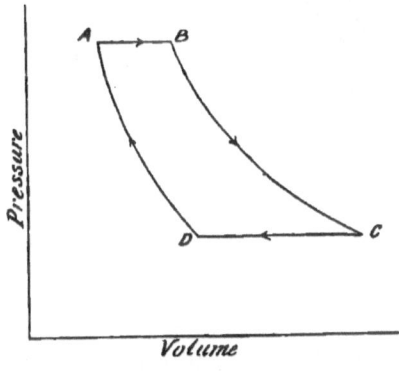

Fig. 43.

c C D d; and, finally, isothermal compression from D to A completes the cycle. The curves A B and C D, fig. 44, follow the equation

$$\phi = Cp \times \log_e \frac{\tau_1}{\tau_2},$$

where Cp = specific heat of 1 lb. air at constant pressure = 0·2375.

The net work done per cycle and per pound of air is the area enclosed by A B C D, and the net heat supplied, exclusive of that received from and given to the regenerator, is b B C c; therefore the efficiency of the cycle is $\dfrac{ABCD}{bBCc}$. But the area a A B b must be equal to the area d D C c; and the work done, or A B C D, is therefore equal to E B C F ; or the efficiency is

$$\frac{EBCF}{bBCc} = \frac{\tau_1 - \tau_2}{\tau_1}$$

as in the case of the Stirling engine.

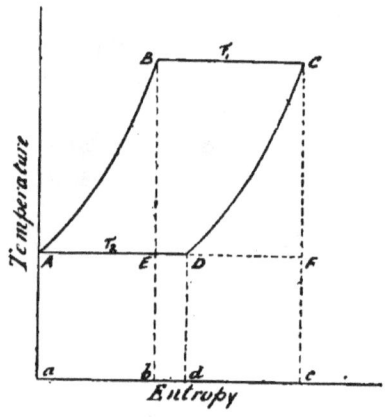

Fig. 44.

33.—Entropy Diagram for Refrigerators.

The cycle of operations in refrigerators is exactly the reverse of that in the Carnot hot-air engine. Instead of taking in heat at a high temperature τ_1, and transforming part of it into work, and rejecting the remainder at a lower temperature τ_2, as in the heat engine, the working substance in the refrigerator receives its heat at the lower temperature τ_2, and discharges it at a higher temperature τ_1, the extra energy required being obtained from external work done on the gas. The theoretically perfect cycle that is reversible is shown in fig. 45 with pressure volume ordinates, and in fig. 46 with temperature entropy ordinates. The first stage of the cycle, A to B, consists of the adiabatic expansion of a certain quantity of air, the temperature falling from τ_1 to τ_2. From B to C the expansion is continued isothermally at constant temperature τ_2, the air receiving heat from the body which it is desired to cool, the amount of heat abstracted being equal to the area E B C F, fig. 46. Compression commences at C, and is at first carried on adiabatically at constant entropy (or isentropically) from C to D, the

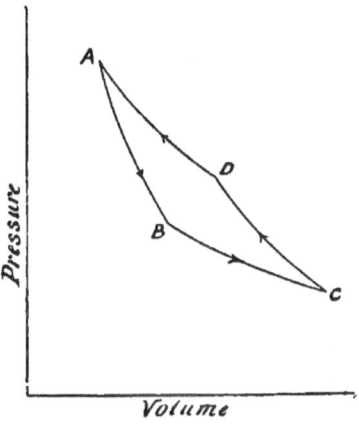

Fig. 45.

temperature rising from τ_2 to τ_1, and is finally completed by isothermal compression from D to A, at constant temperature τ_1, a quantity of heat being rejected to the water jacket equal to F D A E. The heat expended in the process is the equivalent of the work done *on* the gas, and is equal to the area A B C D in both diagrams. The heat absorbed from the substance to be cooled is equal to the rectangle E B C F, fig. 46, and the efficiency, therefore (in its thermo-dynamic sense), is equal to the ratio

$$\frac{EBCF}{ABCD} = \frac{\tau_2}{\tau_1 - \tau_2}.$$

It is thus seen clearly how the efficiency is increased by reducing the difference of temperature between τ_1 and τ_2; and as the ratio

$$\frac{\tau_2}{\tau_1 - \tau_2}$$

may sometimes be greater than unity, it is better known as the "coefficient of performance" (see Howard Lectures, by Professor Ewing, on "The Mechanical Production of Cold," Society of Arts, 1897).

The series of operations in air refrigerators with an open cycle is somewhat different, and is shown in figs. 47 and 48. In this case the air is taken from the cold room and com-

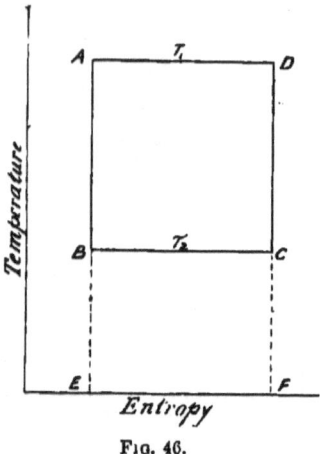

Fig. 46.

pressed adiabatically from A to B. It is then cooled at constant pressure, the temperature falling from B to C,

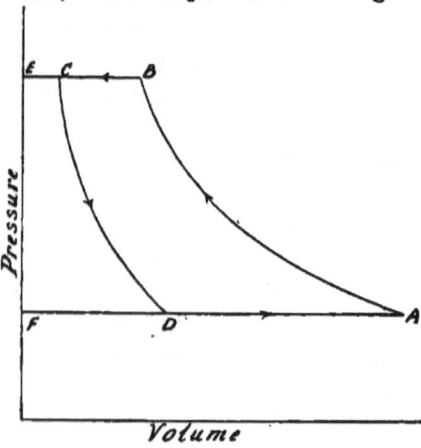

Fig. 47.

fig. 48, and contracting in volume from B to C, fig. 47, after which it is passed into the expansion cylinder, where it expands adiabatically from C to D, and is discharged to the cold room again. The work done *on* the air in the compression cylinder is equal to the area E B A F, fig. 47, or G C B H, fig. 48, and that done *by* the air in the expansion cylinder is equal to E C D F, fig. 47, or G D A H, fig. 48; so that the net external work required is the difference of these two quantities, represented by the area enclosed by A B C D in

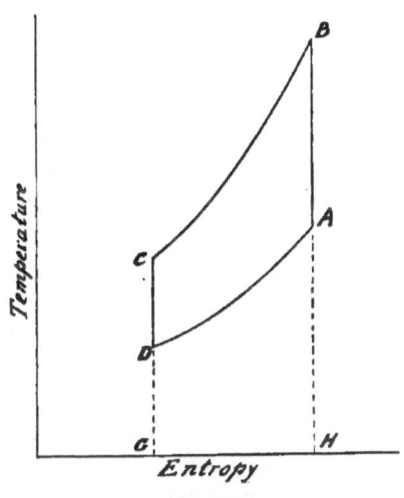

Fig. 48.

both diagrams. The efficiency of the process will be represented by the ratio of the two areas,
$$\frac{ECDF}{EBAF}, \text{ fig. 47};$$
but, as A B and C D are similar adiabatic curves, this will be equal to the ratio
$$\frac{EC}{EB} \text{ or } \frac{FD}{FA}$$

APPENDIX.

Table of the Weight of Dry Saturated Steam.

Having been engaged for some time past in carrying out and calculating the results of steam engine and boiler trials, the author has keenly felt the want of a more complete table of the properties of saturated steam than those hitherto published; one which would give the temperature, density, and total and latent heats for very slight intervals of pressure, in order to save the inconvenience of so much interpolation. In furtherance of this object, he compiled some time ago a new Table of the Density of Dry Saturated Steam, taking the well-known table of Prof. Dwelshauvers Dery, of Liege, as a basis, and extending it so as to give directly the density for every $\frac{1}{10}$ lb. difference of pressure, with a list of additional weights which may be added to give the density for every $\frac{1}{100}$ lb. difference in pressure; the latter, of course, only being used when extreme accuracy is required. It is published now for the first time, with the hope that it may be of some service to others engaged in similar work.

APPENDIX.

WEIGHT OF DRY SATURATED STEAM.

Pressure. Lbs. per sq. in. abs.	Weight. Lbs. per cubic f..ot.	Δ Weight per 1 lb. pres.	Weight in pounds per cubic foot for each $\tfrac{1}{10}$ lb. pressure.										Δ Weight for each $\tfrac{1}{100}$ lb. pressure.									
			0·1	0·2	0·3	0·4	0·5	0·6	0·7	0·8	0·9	·01	·02	·03	·04	·05	·06	·07	·08	·09
1	·00299	·00278	·00327	·00355	·00383	·00411	·00439	·00467	·00494	·00522	·00550	3	6	8	11	14	17	20	22	25
2	·00577	·00271	·00604	·00631	·00658	·00686	·00713	·00740	·00767	·00794	·00821		6	8	11	14	16	19	22	24
3	·00848	·00264	·00874	·00900	·00927	·00953	·00980	·01007	·01033	·01059	·01086	3	6	8	10	13	16	18	21	23
4	·01112	·00261	·01138	·01164	·01190	·01217	·01243	·01269	·01295	·01321	·01347	3	5	8	10	13	16	18	21	23
5	·01373	·00258	·01399	·01424	·01450	·01476	·01502	·01528	·01554	·01579	·01605	3	5	8	10	13	15	18	20	23
6	·01631	·00256	·01657	·01683	·01708	·01734	·01760	·01785	·01811	·01836	·01861	3	5	8	10	12	15	17	20	22
7	·01887	·00253	·01912	·01938	·01963	·01988	·02013	·02039	·02064	·02089	·02114	3	5	7						
8	·02140	·00251	·02165	·02190	·02215	·02240	·02265	·02291	·02316	·02341	·02366	2	5	7	10	12	14	17	20	22
9	·02391	·00250	·02416	·02441	·02466	·02491	·02516	·02541	·02566	·02591	·02316	2	5							
10	·02641	·00248	·02666	·02690	·02715	·02740	·02764	·02789	·02814	·02839	·02864									
11	·02889	·00247	·02914	·02939	·02964	·02988	·03013	·03038	·03062	·03087	·03111									
12	·03136	·00245	·03160	·03185	·03209	·03234	·03258	·03283	·03307	·03332	·03356									
13	·03381	·00244	·03405	·03430	·03454	·03478	·03503	·03527	·03552	·03576	·03601									
14	·03625	·00243	·03649	·03673	·03698	·03722	·03746	·03770	·03795	·03819	·03843									
15	·03868	·00242	·03892	·03916	·03940	·03965	·03989	·04013	·04037	·04061	·04085									

APPENDIX. 111

									2	5	7	10	12	14	17	19	22
16	·04110	·00241	·04134	·04158	·04182	·04206	·04230	·04255	·04279	·04308	·04327
17	·04351	·00240	·04376	·04399	·04423	·04447	·04471	·04495	·04519	·04543	·04567
18	·04591	·00240	·04615	·04639	·04663	·04687	·04711	·04735	·04759	·04783	·04807
19	·04831	·00239	·04855	·04879	·04902	·04926	·04950	·04974	·04998	·05022	·05046

									2	5	7	10	12	14	16	19	21
20	·05070	·00238	·05094	·05117	·05141	·05165	·05183	·05212	·05236	·05260	·05284
21	·05308	·00237	·05332	·05355	·05379	·05403	·05426	·05450	·05473	·05497	·05521
22	·05545	·00237	·05569	·05592	·05616	·05640	·05663	·05687	·05711	·05734	·05758
23	·05782	·00236	·05806	·05829	·05853	·05876	·05900	·05923	·05917	·05970	·05994
24	·06018	·00235	·06041	·06085	·06083	·06112	·06135	·06159	·06182	·06206	·06239
25	·06253	·00234	·06276	·06299	·06323	·06346	·06370	·06393	·06416	·06440	·06463
26	·06487	·00234	·06510	·06533	·06557	·06580	·06603	·06627	·06650	·06673	·06697
27	·06721	·00234	·06744	·06768	·06791	·06814	·06838	·06861	·06884	·06908	·06931
28	·06955	·00233	·06978	·07001	·07025	·07046	·07071	·07094	·07118	·07141	·07164
29	·07188	·00232	·07211	·07234	·07258	·07281	·07304	·07327	·07350	·07373	·07397

									2	5	7	9	12	14	16	19	21
30	·07430	·00232	·07443	·07466	·07490	·07513	·07536	·07559	·07582	·07606	·07639
31	·07652	·00332	·07675	·07698	·07722	·07745	·07768	·07791	·07814	·07837	·07860
32	·07884	·00231	·07907	·07931	·07954	·07977	·08000	·08023	·08046	·08069	·05092

Continued on next page.

APPENDIX.

WEIGHT OF DRY SATURATED STEAM—continued.

Pressure. Lbs. per sq. in. abs.	Weight. Lbs. per cubic foot.	Δ Weight per 1 lb. press.	Weight in pounds per cubic foot for each $\frac{1}{10}$ lb. pressure.										Δ Weight for each $\frac{1}{100}$ lb. pressure.									
			0·1	0·2	0·3	0·4	0·5	0·6	0·7	0·8	0·9	·01	·02	·03	·04	·05	·06	·07	·08	·09
33	·08115	·00231	·08138	·08162	·08185	·08208	·08231	·08254	·08277	·08300	·08323	2	5	7	9	11	14	16	18	21
34	·08346	·00230	·08369	·08392	·08415	·08438	·08461	·08484	·08507	·08530	·08553									
35	·08576	·00230	·08599	·08622	·08645	·08668	·08691	·08714	·08737	·08760	·08783									
36	·08806	·00229	·08829	·08852	·08875	·08897	·08920	·08943	·08966	·08989	·09012									
37	·09035	·00229	·09058	·09081	·09104	·09127	·09150	·09173	·09195	·09218	·09241									
38	·09264	·00229	·09287	·09310	·09332	·09355	·09378	·09401	·09424	·09447	·09470									
39	·09493	·00228	·09515	·09538	·09560	·09583	·09606	·09629	·09652	·09675	·09698									
40	·09721	·00228	·09744	·09767	·09789	·09812	·09835	·09858	·09881	·09903	·09926	2	5	7	9	11	14	16	18	20
41	·09949	·00228	·09972	·09994	·10017	·10040	·10062	·10085	·10108	·10131	·10154									
42	·10177	·00227	·10200	·10222	·10245	·10268	·10290	·10313	·10336	·10358	·10381									
43	·10404	·00227	·10427	·10449	·10472	·10495	·10518	·10540	·10563	·10586	·10608									
44	·10631	·00226	·10654	·10676	·10699	·10721	·10744	·10767	·10789	·10812	·10835									
45	·10857	·00226	·10880	·10902	·10925	·10947	·10970	·10992	·11015	·11037	·11060									
46	·11083	·00226	·11106	·11128	·11151	·11173	·11196	·11218	·11241	·11263	·11286									
47	·11309	·00226	·11332	·11355	·11377	·11400	·11422	·11445	·11467	·11490	·11512									

APPENDIX. 113

No.		D	2	4	7	9	11	13	16	18	20
48	·11535	·00225	·11557	·11580	·11602	·11625	·11647	·11670	·11692	·11715	·11737
49	·11760	·00225	·11782	·11805	·11827	·11850	·11872	·11895	·11917	·11940	·11962
50	·11985	·00225	·12007	·12030	·12052	·12075	·12097	·12120	·12142	·12165	·12187
51	·12210	·00224	·12232	·12255	·12277	·12300	·12322	·12345	·12367	·12390	·12412
52	·12434	·00224	·12456	·12479	·12501	·12523	·12546	·12568	·12591	·12613	·12636
53	·12658	·00224	·12680	·12703	·12725	·12748	·12770	·12793	·12815	·12838	·12860
54	·12882	·00224	·12904	·12927	·12949	·12972	·12994	·13017	·13039	·13061	·13083
55	·13105	·00223	·13127	·13149	·13172	·13194	·13216	·13239	·13261	·13284	·13306
56	·13328	·00223	·13350	·13372	·13395	·13417	·13439	·13461	·13483	·13506	·13528
57	·13551	·00223	·13573	·13595	·13618	·13640	·13662	·13684	·13707	·13729	·13751
58	·13774	·00223	·13796	·13818	·13841	·13863	·13885	·13908	·13920	·13932	·13975
59	·13997	·00222	·14019	·14041	·14064	·14086	·14108	·14130	·14153	·14175	·14197
60	·14219	·00222	·14241	·14263	·14286	·14308	·14330	·14352	·14375	·14397	·14419
61	·14441	·00222	·14463	·14486	·14508	·14530	·14552	·14574	·14596	·14619	·14641
62	·14663	·00222	·14685	·14708	·14730	·14752	·14774	·14796	·14818	·14841	·14863
63	·14885	·00222	·14907	·14930	·14952	·14974	·14996	·15018	·15041	·15063	·15085
64	·15107	·00222	·15129	·15152	·15174	·15196	·15218	·15240	·15263	·15285	·15307

Continued on next page.

APPENDIX.

WEIGHT OF DRY SATURATED STEAM—continued.

Pressure, Lbs. per sq. in. abs.	Weight, Lbs. per cubic foot.	Δ Weight per lb. press.	Weight in pounds per cubic foot for each $\frac{1}{10}$ lb. pressure.										Δ Weight for each $\frac{1}{100}$ lb. pressure.								
			0·1	0·2	0·3	0·4	0·5	0·6	0·7	0·8	0·9	·01	·02	·03	·04	·05	·06	·07	·08	·09
65	·15329	·00222	·15351	·15374	·15396	·15418	·15440	·15462	·15485	·15507	·15529	2	4	7	9	11	13	16	18	20
66	·15551	·00221	·15573	·15595	·15617	·15640	·15662	·15684	·15706	·15728	·15750									
67	·15772	·00220	·15794	·15816	·15838	·15860	·15882	·15904	·15926	·15948	·15970									
68	·15992	·00220	·16014	·16036	·16058	·16080	·16102	·16124	·16146	·16168	·16190									
69	·16212	·00220	·16234	·16256	·16278	·16300	·16322	·16344	·16366	·16388	·16410									
70	·16432	·00220	·16454	·16476	·16498	·16520	·16542	·16564	·16586	·16608	·16630	2	4	7	9	11	13	15	18	20
71	·16652	·00219	·16674	·16696	·16718	·16740	·16762	·16783	·16805	·16827	·16849									
72	·16871	·00219	·16893	·16915	·16937	·16959	·16981	·17002	·17024	·17046	·17068									
73	·17090	·00219	·17112	·17134	·17156	·17178	·17200	·17221	·17243	·17265	·17287									
74	·17309	·00219	·17331	·17353	·17375	·17397	·17418	·17440	·17462	·17484	·17506									
75	·17528	·00219	·17550	·17572	·17594	·17615	·17637	·17659	·17681	·17703	·17725									
76	·17747	·00219	·17769	·17791	·17813	·17834	·17856	·17878	·17900	·17922	·17944									
77	·17966	·00219	·17988	·18010	·18031	·18053	·18075	·18097	·18119	·18141	·18163									
78	·18185	·00219	·18207	·18229	·18250	·18272	·18294	·18316	·18338	·18360	·18382									
79	·18404	·00219	·18426	·18448	·18469	·18491	·18513	·18535	·18557	·18579	·18601									

APPENDIX.

						2	4	7	9	11	15	15	17	20
80	.18623	.00218	.18645	.18666	.18688	.18710	.18732	.18754	.18776	.18798	.18819
81	.18841	.00218	.18863	.18885	.18906	.18928	.18950	.18972	.18994	.19015	.19037
82	.19059	.00218	.19081	.19103	.19124	.19146	.19168	.19190	.19211	.19233	.19255
83	.19277	.00218	.19299	.19320	.19342	.19364	.19386	.19407	.19429	.19451	.19473
84	.19495	.00218	.19517	.19538	.19560	.19582	.19604	.19625	.19647	.19669	.19691
85	.19713	.00218	.19734	.19756	.19778	.19800	.19822	.19843	.19865	.19887	.19909
86	.19931	.00217	.19952	.19974	.19996	.20018	.20040	.20061	.20083	.20105	.20127
87	.20148	.00217	.20170	.20192	.20213	.20235	.20257	.20278	.20300	.20322	.20343
88	.20365	.00217	.20387	.20419	.20430	.20452	.20474	.20495	.20517	.20539	.20560
89	.20582	.00217	.20604	.20625	.20647	.20669	.20691	.20712	.20734	.20756	.20777
90	.20799	.00217	.20821	.20842	.20864	.20886	.20907	.20929	.20951	.20972	.20994
91	.21016	.00216	.21038	.21059	.21081	.21103	.21124	.21146	.21167	.21189	.21211
92	.21232	.00216	.21254	.21275	.21297	.21319	.21340	.21362	.21383	.21405	.21427
93	.21448	.00216	.21470	.21491	.21513	.21535	.21556	.21578	.21599	.21621	.21643
94	.21664	.00216	.21686	.21707	.21729	.21750	.21772	.21793	.21815	.21837	.21858
95	.21880	.00215	.21901	.21923	.21944	.21966	.21987	.22009	.22030	.22052	.22073
96	.22095	.00215	.22116	.22138	.22159	.22181	.22202	.22224	.22245	.22267	.22288
97	.22310	.00215	.22331	.22353	.22374	.22396	.22417	.22439	.22460	.22482	.22503

Continued on next page.

APPENDIX.

WEIGHT OF DRY SATURATED STEAM—continued.

Pressure. Lbs. per sq. in. abs.	Weight. Lbs. per cubic foot.	Δ Weight per 1 lb. press.	Weight in pounds per cubic foot for each 1/10 lb. pressure.										Δ Weight for each 1/100 lb. pressure.									
			0·1	0·2	0·3	0·4	0·5	0·6	0·7	0·8	0·9	·01	·02	·03	·04	·05	·06	·07	·08	·09
98	·22525	·00215	·22540	·22568	·22589	·22611	·22632	·22654	·22675	·22697	·22718		2							
99	·22740	·00215	·22761	·22783	·22804	·22826	·22847	·22869	·22890	·22912	·22933									
100	·22955	·00215	·22976	·22998	·23019	·23041	·23062	·23084	·23105	·23127	·23148									
101	·23170	·00215	·23191	·23213	·23234	·23256	·23277	·23299	·23320	·23342	·23363									
102	·23385	·00215	·23406	·23428	·23449	·23471	·23492	·23514	·23535	·23557	·23578									
103	·23600	·00215	·23621	·23643	·23664	·23686	·23707	·23729	·23750	·23772	·23793									
104	·23815	·00215	·23836	·23858	·23879	·23901	·23922	·23944	·23965	·23987	·24008		4	6	9	11	13	15	17	19
105	·24030	·00215	·24051	·24073	·24094	·24116	·24137	·24159	·24180	·24202	·24223									
106	·24245	·00215	·24266	·24288	·24309	·24331	·24352	·24374	·24395	·24417	·24438									
107	·24460	·00214	·24481	·24503	·24524	·24546	·24567	·24588	·24610	·24631	·24652									
108	·24674	·00213	·24695	·24716	·24738	·24759	·24780	·24802	·24823	·24844	·24866									
109	·24887	·00213	·24908	·24929	·24951	·24972	·24993	·25015	·25036	·25057	·25079									
110	·25100	·00214	·25121	·25143	·25164	·25185	·25207	·25228	·25249	·25271	·25292									
111	·25314	·00214	·25335	·25357	·25378	·25399	·25421	·25442	·25464	·25485	·25506									

APPENDIX. 117

112	·25528	·00214	·25549	·25571	·25592	·25613	·25635	·25656	·25678	·25699	·25720	2	4	6	9	11	13	15	17	19
113	·25742	·00214	·25763	·25785	·25806	·25827	·25849	·25870	·25892	·25913	·25934									
114	·25956	·00214	·25977	·25999	·26020	·26041	·26063	·26084	·26106	·26127	·26148									
115	·26170	·00214	·26192	·26213	·26235	·26257	·26278	·26299	·26321	·26342	·26363	2	4	6	9	11	13	15	17	19
116	·26384	·00214	·26405	·26427	·26448	·26469	·26491	·26512	·26534	·26555	·26576									
117	·26598	·00214	·26619	·26641	·26662	·26683	·26705	·26726	·26748	·26769	·26790									
118	·26812	·00213	·26833	·26855	·26876	·26897	·26919	·26940	·26961	·26983	·27004									
119	27025	·00213	·27046	·27068	·27089	·27110	·27132	·27153	·27174	·27196	·27217									
120	·27238	·00213	·27259	·27281	·27302	·27323	·27345	·27366	·27387	·27408	·27430	2	4	6	8	11	13	15	17	19
121	·27451	·00212	·27472	·27494	·27515	·27536	·27557	·27579	·27600	·27621	·27643									
122	·27663	·00212	·27684	·27705	·27727	·27748	·27769	·27791	·27812	·27833	·27854									
123	·27875	·00212	·27896	·27917	·27939	·27960	·27981	·28002	·28024	·28045	·28066									
124	·28087	·00212	·28108	·28130	·28151	·28172	·28193	·28214	·28235	·28256	·28278									
125	·28299	·00212	·28320	·28341	·28363	·28384	·28405	·28426	·28448	·28469	·28490									
126	·28511	·00212	·28532	·28554	·28575	·28596	·28617	·28638	·28660	·28681	·28702									
127	·28723	·00212	·28744	·28766	·28787	·28808	·28829	·28850	·28872	·28893	·28914									
128	·28935	·00212	·28956	·28978	·28999	·29020	·29041	·29063	·29084	·29105	·29126									
129	·29147	·00212	·29168	·29190	·29211	·29232	·29253	·29275	·29296	·29317	·29338									

Continued on next page.

APPENDIX.

WEIGHT OF DRY SATURATED STEAM—continued.

Pressure. Lbs. per sq. in. abs.	Weight Lbs. per cubic foot	Δ Weight per 1 lb. press.	0·1	0·2	0·3	0·4	0·5	0·6	0·7	0·8	0·9	Δ Weight for each 1/10 lb. pressure.									
												0·1	0·2	0·3	0·4	0·5	0·6	0·7	0·8	0·9
130	·29359	·00212	·29380	·29402	·29423	·29444	·29465	·29487	·29508	·29529	·29550	2	4	6	8	11	13	15	17	19
131	·29571	·00212	·29592	·29614	·29635	·29656	·29677	·29699	·29720	·29741	·29762									
132	·29783	·00212	·29804	·29826	·29847	·29868	·29889	·29910	·29932	·29953	·29974									
133	·29995	·00212	·30016	·30037	·30059	·30080	·30101	·30122	·30143	·30165	·30186									
134	·30207	·00212	·30228	·30249	·30270	·30292	·30313	·30334	·30355	·30376	·30398									
135	·30419	·00211	·30440	·30461	·30482	·30503	·30525	·30546	·30567	·30588	·30609									
136	·30630	·00211	·30651	·30672	·30693	·30714	·30736	·30757	·30778	·30799	·30820									
137	·30841	·00211	·30862	·30883	·30904	·30925	·30947	·30968	·30989	·31010	·31031									
138	·31052	·00211	·31073	·31094	·31115	·31136	·31157	·31179	·31200	·31221	·31242									
139	·31263	·00211	·31284	·31305	·31326	·31347	·31368	·31390	·31411	·31432	·31453									
140	·31474	·00211	·31495	·31516	·31537	·31558	·31579	·31600	·31622	·31643	·31664									
141	·31685	·00211	·31706	·31727	·31748	·31769	·31790	·31812	·31833	·31854	·31875									
142	·31896	·00211	·31917	·31938	·31959	·31980	·32001	·32022	·32043	·32065	·32086									
143	·32107	·00211	·32128	·32149	·32170	·32191	·32212	·32233	·32254	·32275	·32297									
144	·32318	·00211	·32339	·32360	·32381	·32402	·32423	·32444	·32465	·32486	·32508									

APPENDIX. 119

							2	4	6	8	10	13	15	17	19
145	·32529	·00210	·32550	·32571	·32592	·32613	·32634	·32655	·32676	·32697	·32718				
146	·32739	·00210	·32760	·32781	·32802	·32823	·32844	·32865	·32886	·32907	·32928				
147	·32949	·00210	·32970	·32991	·33012	·33033	·33054	·33075	·33096	·33117	·33138				
148	·33159	·00210	·33180	·33201	·33222	·33243	·33264	·33285	·33306	·33327	·33348				
149	·33369	·00210	·33390	·33411	·33432	·33453	·33474	·33495	·33516	·33537	·33558				
150	·33579	·00210	·33600	·33621	·33642	·33663	·33684	·33705	·33726	·33747	·33768				
151	·33789	·00209	·33810	·33831	·33852	·33873	·33894	·33915	·33936	·33957	·33977				
152	·33998	·00209	·34019	·34040	·34061	·34082	·34103	·34124	·34145	·34165	·34186				
153	·34207	·00209	·34228	·34249	·34270	·34291	·34312	·34333	·34353	·34374	·34395				
154	·34416	·00209	·34437	·34458	·34479	·34500	·34521	·34541	·34562	·34583	·34604				
155	·34625	·00209	·34646	·34667	·34688	·34709	·34729	·34750	·34771	·34792	·34813				
156	·34834	·00209	·34855	·34876	·34897	·34918	·34933	·34959	·34980	·35001	·35022				
157	·35043	·00209	·35064	·35085	·35106	·35126	·35147	·35168	·35189	·35210	·35231				
158	·35252	·00209	·35273	·35294	·35315	·35335	·35356	·35377	·35393	·35419	·35440				
159	·35461	·00209	·35482	·35503	·35523	·35544	·35565	·35586	·35607	·35628	·35649				
160	·35670	·00209	·35691	·35712	·35732	·35753	·35774	·35795	·35816	·35837	·35858				
161	·35879	·00209	·35900	·35920	·35941	·35962	·35983	·36004	·36025	·36046	·36067				

Continued on next page.

APPENDIX.

WEIGHT OF DRY SATURATED STEAM—continued.

Pressure. Lbs per sq. in abs.	Weight. Lbs. per cubic foot.	Δ Weight per 1 lb. press.	Weight in pounds per cubic foot for each 1/10 lb. pressure.									Δ Weight for each 1/100 lb. pressure.								
			0·1	0·2	0·3	0·4	0·5	0·6	0·7	0·8	0·9	·01	·02	·03	·04	·05	·06	·07	·08	·09
162	·36036	·00208	·36108	·36129	·36150	·36171	·36192	·36213	·36234	·36254	·36275	2	4	6	8	10	12	15	17	19
163	·36296	·00208	·36317	·36338	·36358	·36379	·36400	·36421	·36442	·36462	·36483									
164	·36504	·00208	·36525	·36546	·36566	·36587	·36608	·36629	·36650	·36670	·36691									
165	·36712	·00208	·36733	·36754	·36774	·36795	·36816	·36837	·36857	·36878	·36899									
166	·36920	·00203	·36941	·36962	·36982	·37003	·37024	·37045	·37066	·37086	·37107									
167	·37128	·00208	·37149	·37170	·37190	·37211	·37232	·37253	·37274	·37294	·37315									
168	·37336	·00208	·37357	·37378	·37398	·37419	·37440	·37461	·37482	·37502	·37523									
169	·37544	·00208	·37565	·37586	·37606	·37627	·37648	·37660	·37690	·37710	·37731									
170	·37752	·00208	·37773	·37794	·37814	·37835	·37856	·37877	·37898	·37918	·37930	2	4	6	8	10	12	15	17	19
171	·37960	·00208	·37981	·38002	·38022	·38043	·38064	·38085	·38105	·38126	·38147									
172	·38168	·00208	·38189	·38210	·38230	·38251	·38272	·38293	·38314	·38334	·38355									
173	·38376	·00208	·38397	·38418	·38438	·38459	·38480	·38501	·38522	·38542	·38563									
174	·38584	·00208	·38605	·38626	·38646	·38667	·38688	·38709	·38730	·38750	·38771									
175	·38792	·00208	·38813	·38834	·38854	·38875	·38896	·38917	·38938	·38958	·38979									
176	·39000	·00208	·39021	·39012	·39062	·39083	·39104	·39125	·39146	·39166	·39187									

APPENDIX.

			2	4	6	8	10	12	15	17	19
177	·39208	·00208	·39229	·39250	·39270	·39291	·39312	·39333	·39354	·39374	·39395
178	·39416	·00208	·39437	·39458	·39478	·39499	·39520	·39541	·39562	·39582	·39603
179	·39624	·00208	·39645	·39666	·39686	·39707	·39728	·39749	·39770	·39790	·39811
180	·39832	·00208	·39853	·39874	·39894	·39915	·39936	·39957	·39978	·39998	·40019
181	·40040	·00208	·40061	·40082	·40102	·40123	·40144	·40165	·40186	·40206	·40227
182	·40249	·00208	·40269	·40290	·40310	·40331	·40352	·40373	·40394	·40414	·40435
183	·40456	·00208	·40477	·40498	·40518	·40539	·40560	·40581	·40602	·40622	·40643
184	·40664	·00208	·40685	·40706	·40726	·40747	·40768	·40789	·40810	·40830	·40851
185	·40872	·00208	·40893	·40914	·40934	·40955	·40976	·40997	·41018	·41038	·41059
186	·41080	·00208	·41101	·41122	·41142	·41163	·41184	·41205	·41226	·41246	·41267
187	·41288	·00208	·41309	·41330	·41350	·41371	·41392	·41413	·41434	·41454	·41475
188	·41496	·00208	·41517	·41538	·41558	·41579	·41600	·41621	·41642	·41662	·41683
189	·41704	·00208	·41725	·41746	·41766	·41787	·41808	·41829	·41850	·41870	·41991
190	·41912	·00206	·41933	·41954	·41974	·41995	·42016	·42037	·42058	·42078	·42099
191	·42120	·00206	·42141	·42162	·42182	·42203	·42224	·42245	·42266	·42286	·42307
192	·42328	·00208	·42349	·42370	·42390	·42411	·42432	·42453	·42474	·42494	·42515
193	·42536	·00208	·42551	·42578	·42598	·42619	·42640	·42661	·42682	·42702	·42723

Continued on next page.

APPENDIX.

WEIGHT OF DRY SATURATED STEAM—continued.

Pressure. Lbs. per sq. in. abs.	Weight. Lbs. per cubic foot.	Δ Weight per 1 lb. press.	Weight in pounds per cubic foot for each 1/10 lb. pressure.										Δ Weight for each 1/100 lb. pressure.									
			0·1	0·2	0·3	0·4	0·5	0·6	0·7	0·8	0·9	·01	·02	·03	·04	·05	·06	·07	·08	·09
194	·42744	·00208	·42765	·42783	·42806	·42827	·42848	·42869	·42890	·42910	·42931	2	4	6	8	10	12	15	17	19
195	·42952	·00208	·42973	·42994	·43014	·43035	·43056	·43077	·43098	·43118	·43139									
196	·43160	·00208	·43181	·43202	·43222	·43243	·43264	·43285	·43306	·43326	·43347									
197	·43368	·00208	·43389	·43410	·43430	·43451	·43472	·43493	·43514	·43534	·43555									
198	·43576	·00208	·43597	·43617	·43638	·43659	·43680	·43701	·43721	·43742	·43763									
199	·43784	·00208	·43805	·43825	·43846	·43867	·43888	·43908	·43929	·43950	·43971									
200	·43992	·00207	·44013	·44033	·44054	·44075	·44095	·44116	·44137	·44158	·44178	2	4	6	8	10	12	14	17	19
201	·44199	·00207	·44220	·44240	·44261	·44282	·44303	·44323	·44344	·44365	·44386									
202	·44406	·00207	·44427	·44448	·44468	·44489	·44510	·44530	·44550	·44571	·44592									
203	·44613	·00207	·44634	·44654	·44675	·44696	·44716	·44737	·44758	·44778	·44799									
204	·44820	·00206	·44841	·44861	·44882	·44903	·44923	·44944	·44964	·44985	·45006									
205	·45026	·00206	·45047	·45067	·45088	·45109	·45129	·45150	·45170	·45191	·45212	2	4	6	8	10	12	14	16	19
206	·45232	·00206	·45253	·45273	·45294	·45315	·45335	·45356	·45376	·45397	·45418									
207	·45438	·00205	·45459	·45479	·45500	·45520	·45541	·45561	·45582	·45602	·45623									

APPENDIX. 123

			2	4	6	8	10	12	14	16	18
208	·45643	·00205	·45664	·45684	·45705	·45725	·45746	·45766	·45787	·45807	·45828
209	·45848	·00205	·45869	·45889	·45910	·45930	·45951	·45971	·45992	·46012	·46032
210	·46053	·00204	·46073	·46094	·46114	·46135	·46155	·46175	·46196	·46216	·46237
211	·46257	·00204	·46277	·46293	·46318	·46339	·46359	·46379	·46400	·46420	·46441
212	·46461	·00204	·46481	·46502	·46522	·46543	·46563	·46583	·46604	·46624	·46614
213	·46665	·00204	·46685	·46705	·46726	·46746	·46766	·46787	·46807	·46828	·46848
214	·46869	·00203	·46869	·46910	·46930	·46950	·46971	·46991	·47011	·47032	·47052
215	·47072	·00203	·47092	·47113	·47133	·47153	·47174	·47194	·47214	·47234	·47255
216	·47275	·00202	·47295	·47315	·47336	·47356	·47376	·47396	·47416	·47437	·47457
217	·47477	·00202	·47497	·47517	·47538	·47558	·47578	·47508	·47618	·47639	·47659
218	·47679	·00202	·47699	·47720	·47740	·47760	·47780	·47800	·47820	·47841	·47861
219	·47881	·00202	·47901	·47922	·47942	·47962	·47982	·48002	·49022	·45043	·48063

INDEX.

	PAGE
Adiabatic Expansion Curve	32
Adiabatic Expansion Condensation	35
Adiabatic Expansion of Wet Steam	37
Air Calculations, Gas Engine	83
Air Engines, Diagrams for	101, 104
Aquene Curve	8
Areas of Diagrams, Comparison of	46
Balance Sheet, Heat, Steam Engine	72
Balance Sheet, Heat, Gas Engine	96
Boulvin's (Professor) Complete Entropy Diagram	28
Carnot Cycle	4, 34
Chart, Theta-Phi	2
Clearance Volumes	21
Clearance Surfaces	54
Coefficient of Performance	106
Comparison of Areas	46
Complete Entropy Diagram	28
Compound Engine, Diagram for	23
Compounding, Effect of	59
Condensation Coefficient	59
Condensation during Adiabatic Expansion	36
Condensation during Expansion	44
Condensation, Initial	49, 56, 63
Constant Volume Curves	15, 92
Constant Temperature Lines	30
Conversion of Indicator Diagram	21
Cotterill, Professor, on Steam Engine	19, 30
Critical Temperature	14
Cut-off, Most Economical	61
Density of Steam	22, 109
Diesel Oil Motor, Professor Schröter's Test	97
Diesel Oil Motor, Temperature Calculations	98
Diesel Oil Motor, Entropy Diagram	99
Donkin, Mr. B., on "Experiments on Small Vertical Engine"	49, 57
Donkin, Mr. B., on "Gas, Air, and Oil Engines"	81
Dryness Fraction, Calculation	21
Dryness Fraction, Comparisons of	56
Efficiency, Thermal, Steam Engine	51, 69
Efficiency, Thermal, Stirling's Air Engine	103
Efficiency, Thermal, Refrigerators	106, 108
Engine, Compound, Diagrams for	23
Engine, Single-cylinder, Diagrams for	46
Engine, Triple-expansion, Diagrams for	39

INDEX.

	PAGE
Entropy	3
Entropy of Water	7, 10
Entropy of Steam	11
Entropy of Superheated Steam	46
Entropy of Gases	74
Entropy Diagram	5
Entropy Diagram for Ice, Water, and Steam	11
Entropy Diagram for Gas Engine	87
Entropy Diagram for Oil Engine	97
Entropy Diagram for Air Engines	101, 104
Ericsson's Air Engine	104
Ewing, Professor, on "Mechanical Production of Cold"	104
Exchanges of Heat	68
Exhaust Period, Gas Engine	91
Exhaust Products, Gas Engine	85
Exhaust Waste, Gas Engine	94
Expansion, Adiabatic	32, 35
Expansion, Most Economical	61
Expansion Period, Gas Engine	90
Gas, Coal, Analysis	84
Gas, Coal, Specific Heat	84
Gases, Entropy of	73
Gases, Specific Heat of	73
Gas Engine, Theoretical	76
Gas Engine, Temperature Calculations	78
Gas Engine, Actual, 7 Horse Power	81
Gas Engine, Ideal Diagram	87
Gas Engine, Corrected Diagram	89
Gray, Mr. M. F., on "Theta-Phi Chart"	1
Heat Balance Sheet, Steam Engine	72
Heat Balance Sheet, Gas Engine	96
Heat, Latent	6
Heat Losses, Steam Engine	39
Heat Losses, Measurement of	68
Heat Losses, Gas Engine	94
Heat Recovery Lines	66
Heat Weight	3
High-pressure Cylinder, $\theta \phi$ Diagram	25
Indicator Diagrams Compared with $\theta \phi$	3
Indicator Diagrams Converted to $\theta \phi$	21
Initial Condensation	49, 51, 63
Introduction of $\theta \phi$ Diagrams	1
Jacketing, Steam, Effect of	39
Latent Heat	6
Logarithmic Curves	79
Longridge Mr. M., on "Trials of a Compound Engine"	61
Losses of Heat	39, 68, 94
Low-pressure Cylinder, $\theta \phi$ Diagram	26
Main Valve, Passage in	27
Mixture in Gas-engine Cylinder	82
Oil Engines, $\theta \phi$ Diagram for	97
Priming Water	63

INDEX.

	PAGE
Quality Curves	21
Radiation, Gas-engine	94
Re-evaporation during Expansion	44
Refrigerators, Closed Cycle	105
Refrigerators, Efficiency	106, 108
Refrigerators, Open Cycle	107
Regenerator, Effect of	103, 104
Ripper, Prof., on "Superheated Steam-engine Trials"	49
Sankey, Captain, on "Marine-engine Trials"	2
Scales for Entropy Diagram	7
Schröter, Prof., on "Diesel Oil Motor"	97
Specific Heat, Water	10
Specific Heat, Superheated Steam	46
Specific Heat, Coal Gas	84
Specific Heat, Gas-engine Mixture	86
Specific Volume, Steam	17
Speed, Effect of	56
Standard Engine of Comparison	69
State Points	35
Steam, Entropy Diagram for	5, 11
Steam Jacketing, Effect of	39
Steam, Superheated, $\theta\phi$ Diagram	46
Steam, Superheated, Effect of	49
Steam, Wet, Expansion of	37
Stirling's Air Engine	101
Superheated Steam, $\theta\phi$ Diagram for	46
Superheated Steam, Entropy of	48
Superheated Steam, Effect of	49
Surfaces, Clearance	54
Temperature of Combustion	94
Temperature, Constant, Lines	80
Temperature, Critical	14
Temperature in Gas Engine	86
Temperature in Oil Engine	98
Theoretical Gas Engine	76
Theoretical Temperature, Gas Engine	94
Thermal Efficiency, Steam Engine	51, 69
Thermal Efficiency of Carnot Cycle	69
Thermal Efficiency of Rankine Cycle	69
Thermal Efficiency of Stirling's Air Engine	103
Thermal Efficiency of Refrigerators	106, 108
Triple-expansion Engine' $\theta\phi$ Diagram	39
Volume Factor	27, 33, 45
Volume, Specific, Steam	17
Wall Action, Steam Engine	52
Wall Action, Gas Engine	94
Water, Entropy of	7, 10
Water, Specific Heat of	10
Weight of Steam	22, 109
Wet Steam, Expansion of	37
Willans, Mr. P. W., on "Non-condensing Steam-engine Trials"	2
Willans, Mr. P. W., on "Condensing Steam-engine Trials"	58, 70
Work Losses	44

Printed by JOHN HEYWOOD, Excelsior Works, Manchester.

Second Edition.
Crown 8vo, cloth, price 4s. 6d. net, post free anywhere.

THE INDICATOR AND ITS DIAGRAMS: WITH CHAPTERS ON ENGINE AND BOILER TESTING. By CHARLES DAY, Wh.Sc. Including a Table of Piston Constants.

Crown 8vo, cloth, price 3s. 6d. net, post free anywhere.

PROBLEMS IN MACHINE DESIGN. For the Use of Students, Draughtsmen, and others. By CHAS. H. INNES, M.A., Lecturer on Engineering at the Rutherford College, Newcastle-on-Tyne.

Crown 8vo, cloth, price 6s. net, post free anywhere.

THE APPLICATION OF GRAPHIC METHODS TO THE DESIGN OF STRUCTURES. Specially prepared for the use of Engineers. Profusely illustrated. By W. W. F. PULLEN, Wh.Sc., Assoc.M.Inst.C.E., M.I.Mech.E.

Crown 8vo, price 3s. 6d. net, post free anywhere.

CENTRIFUGAL PUMPS, TURBINES, & WATER MOTORS: including the Theory and Practice of Hydraulics (specially adapted for engineers). By CHAS. H. INNES, M.A., Lecturer on Engineering at Rutherford College, Newcastle-on-Tyne.

Crown 8vo, cloth, price 3s. net, post free anywhere.

ENGINEERING ESTIMATES AND COST ACCOUNTS. By F. G. BURTON, formerly Secretary and General Manager of the Milford Haven Shipbuilding and Engineering Co. Limited.

Crown 8vo, price 2s. net, post free anywhere.

OPENING BRIDGES. By GEORGE WILSON, M.Sc., Demonstrator and Assistant Lecturer at the University College of South Wales, Cardiff.

Crown 8vo, cloth, price 2s. 6d. net, post free anywhere.

THE NAVAL ENGINEER AND THE COMMAND OF THE SEA. A Story of Naval Administration. By FRANCIS G. BURTON, author of "Engineering Estimates and Cost Accounts," &c.

Crown 8vo, cloth, price 3s. 6d. net, post free anywhere.

GRAPHIC METHODS OF ENGINE DESIGN.
By A. H. BARKER, B.A., B.Sc., Wh.Sc., author of "Graphical Calculus," &c.

Crown 8vo, cloth, price 3s. 6d. net, post free anywhere.

INJECTORS: THEORY, CONSTRUCTION, AND WORKING.
By W. W. F. PULLEN, Wh.Sc., Assoc.M.Inst.C.E., M.Inst.M.E.

Crown 8vo, cloth, price 2s. 6d. net, post free anywhere.

PRACTICAL NOTES ON THE CONSTRUCTION OF CRANES AND LIFTING MACHINERY.
By EDWARD C. R. MARKS, Assoc.M.Inst.C.E., M.I.Mech.E.

Crown 8vo, cloth, price 4s. 6d. net, post free anywhere.

MODERN GAS AND OIL ENGINES.
Profusely illustrated. A full and exhaustive Treatment of the Design, Construction, and Working of Gas and Oil Engines up to date. By FREDERICK GROVER, Assoc.M.Inst.C.E.

Crown 8vo, cloth, price 6s. net, post free anywhere.

HEAT AND HEAT ENGINES.
A Treatise on Thermodynamics as Practically Applied to Heat Motors. Specially written for Engineers. By W. C. POPPLEWELL, M.Sc.

Crown 8vo, cloth, price 5s. net, post free anywhere.

MARINE ENGINEERS: THEIR QUALIFICATIONS AND DUTIES.
With Notes on the Care and Management of Marine Engines, Boilers, Machinery, &c. By E. G. CONSTANTINE, Assoc.M.Inst.C.E., M.I.Mech.E.

Crown 8vo, cloth, price 3s. net, post free anywhere.

THE ABC of the DIFFERENTIAL CALCULUS.
By W. D. WANSBROUGH, author of "Portable Engines," "Proportions and Movement of Slide Valves," &c.

THE TECHNICAL PUBLISHING CO. LIMITED,
31, WHITWORTH STREET, MANCHESTER, ENGLAND;
JOHN HEYWOOD, LONDON AND MANCHESTER;
And all Booksellers.

www.ingramcontent.com/pod-product-compliance
Lightning Source LLC
Chambersburg PA
CBHW022134160426
43197CB00009B/1284